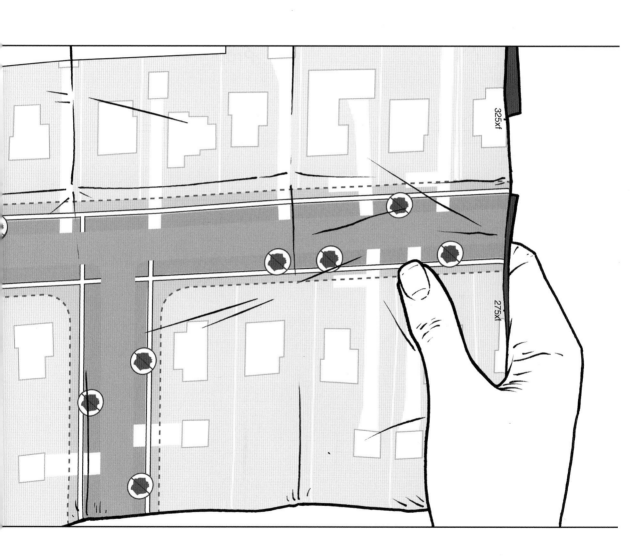

Second Edition

Making MAPS

A Visual Guide to Map Design for GIS

Crestview Rd.

JOHN KRYGIER and DENIS WOOD

THE GUILFORD PRESS
New York London

Printed in the United States of America

This book is printed on acid-free paper.

Last digit is print number: 9 8 7 6 5 4 3 2 1

Library of Congress Cataloging-in-Publication Data

Krygier, John
  Making maps : a visual guide to map design for GIS /
  John Krygier, Denis Wood. — 2nd ed.
      p. cm.
      Includes bibliographical references and index.
      ISBN 978-1-60918-166-6 (pbk.)
    1. Cartography. 2. Geographic information systems.
  I. Wood, Denis. II. Title.
GA105.3.K79 2011
526—dc22
                                          2010040429

It's Time to Make Maps...

People communicate about their places with maps. Less common than talk or writing, maps are made when called for by social circumstances. Jaki and Susan are making maps to protect their neighborhood. Why a map? Because the city used a map. The map unambiguously expresses the city's intentions to widen Crestview Road, drawing from the maps, talk, and text of city planners. If the plan is realized, the city will also use maps to communicate its intentions to surveyors, engineers, contractors, utility companies, and others.

The maps are all of Crestview Road – all of the same place – and the maps are all different. Yet they are all equally good. Different goals call for different maps: the quality of a map is frequently a matter of perspective rather than design. Think of a map *as a kind of statement locating facts.* People will select the facts that make their case. That's what the map is for: to make their case.

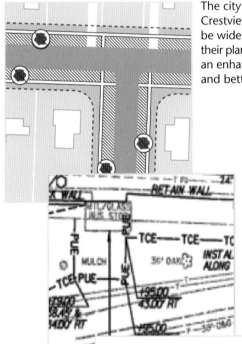

The city's case is that Crestview Road needs to be widened. They present their plan as "a new vision," an enhancement, different and better.

The city communicates to construction firms and utilities with detailed maps, making the case that the planners and engineers have done their work.

Jaki and Susan's case is that widening Crestview Road would be a terrible mistake. Time to make a map!

Making maps, making your case...

Different Goals Call for Different Maps

Jaki and Susan soon realize the plan to widen Crestview is but a piece of a larger plan to redevelop the northern and western suburbs of the city. The key feature of the plan is a connector (in solid black below) proposed to link two major arteries. Different groups create equally effective maps to articulate their different perspectives on the proposed road. Though the maps may seem polemical, isolating the facts each presents is useful in focusing debate.

Goal: keeping costs low. A city map shows that its plan is the shortest and least costly route for the connector. The city's map focuses on moving traffic at the least cost to taxpayers.

Goal: defending neighborhood integrity. An African American community map shows how the connector rubs salt in the wound sustained by the earlier invasion of the arterial highway. The focus of their map is the further destruction of their neighborhood by the proposed connector.

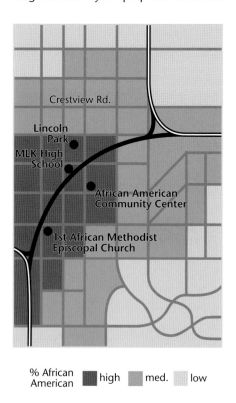

Crestview Rd.

Proposed Connector

Property Values ▮ high ▮ med. ▯ low

Crestview Rd.

Lincoln Park
MLK High School
African American Community Center
1st African Methodist Episcopal Church

% African American ▮ high ▮ med. ▯ low

Goal: maintaining historic continuity. The Society for Historic Preservation's map shows how the connector will affect significant properties in an existing historic district. Their map focuses on the adverse effect on significant properties and on the integrity of the historic district.

Goal: protecting endangered wetlands. An environmental group shows that the connector will violate the city's policy of avoiding road construction in floodplains. The Oberlin Creek watershed, already greatly impacted by over 100 years of urban growth, cannot withstand a further onslaught of development.

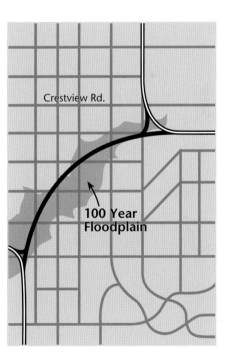

Crestview Rd.

Olmsted's
Lincoln
Park

Oldest Home
in City

Historic
City Hall

Historic
"Shotgun"
Houses

**Oberlin
Historic
District**

% Historic
Buildings high med. low

Crestview Rd.

**100 Year
Floodplain**

Goal: defending their street. Jaki and Susan's first map scales roads to show existing traffic counts. It suggests how much more effective it would be to widen Armitage Avenue, a street already tied into the downtown grid. Their focus is to deflect attention from Crestview Road.

Goal: defeating the connector. Aware of the connector's role in motivating the widening of Crestview, and informed by the maps produced by other groups, Jaki and Susan realize it's less that Crestview needs defending and more that the connector needs defeating: low property values correlate with historic discrimination against African Americans, with older housing, and the floodplain. The connector exploits this nexus: their new map focuses on social and environmental justice. Jaki and Susan work out a "Social and Environmental Justice Sensitivity" metric, taking into account race, history, and environmental factors.

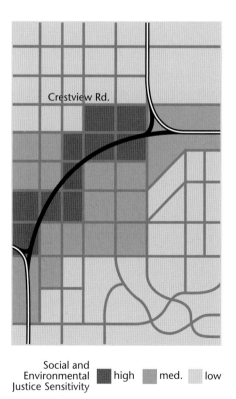

Crestview Rd.

Armitage Ave.

Crestview Rd.

Daily Traffic Counts — high — med. — low

Social and Environmental Justice Sensitivity — high — med. — low

Goal: defeating the connector. Jaki and Susan find interesting information while researching the proposed connector. They change scale and map this part of the story: behind the connector lie quiet negotiations between the state, an international pharmaceutical firm, and well-connected real-estate interests eager to develop farmland to the southwest of the city. The focus now is on the power of lobbyists and back room deals. Jaki and Susan's maps are published with a story on the controversy over the proposed connector in a local independent newspaper.

Goal: going in for the kill. The independent newspaper jumps scale again, mapping the seamy underside of the pharmaceutical firm behind the proposed connector. The focus is on the reach and impact of firms operating on the global scale. Jaki and Susan are astounded that a far-off multinational corporation is behind the threat to Crestview.

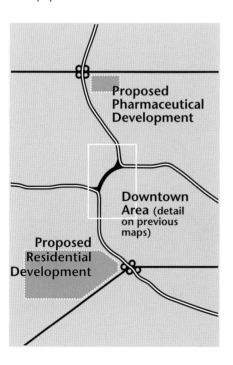

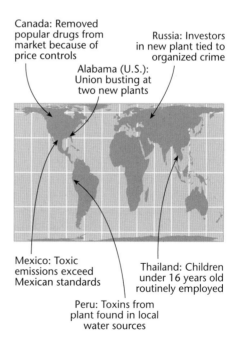

Canada: Removed popular drugs from market because of price controls

Russia: Investors in new plant tied to organized crime

Alabama (U.S.): Union busting at two new plants

Mexico: Toxic emissions exceed Mexican standards

Thailand: Children under 16 years old routinely employed

Peru: Toxins from plant found in local water sources

Different Goals Produce Different Maps

The eight maps involved in this debate over the location of the connector are all good. Each is clear. Each makes its points with accurate data in a way that is easy to read and understand. What makes the maps different is the different purposes each was designed to serve. It is this purpose that drove the selection of facts and these facts that dictated the design and scale. The story continues...

Goal: consider the alternatives. Due to historical, environmental, and social justice concerns with the proposed connector, and the embarrassing newspaper article, the city council asks the city planning department to develop alternatives. When these alternatives are mapped, they raise additional concerns (and maps). Route B, while more costly than A, is cheaper than C (which passes through property owned by influential property developers opposed to the connector). B also has a lower environmental impact and does not adversely affect any organized social groups or business interests.

Goal: seek funds for the proposed bypass. The newly proposed bypass will cost significantly more than the downtown alternative, so the city seeks additional funding. The grant proposals include, among many maps, a map showing the general location of the proposed bypass.

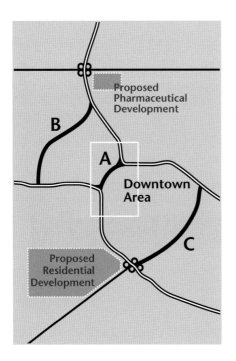

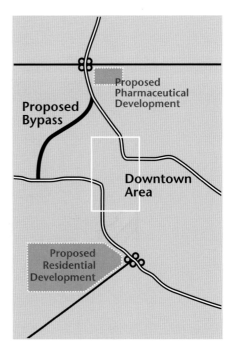

Crestview is saved! Jaki and Susan throw a party to celebrate. They include a map on the party flyer...

Making Maps:
A Visual Guide to Map Design for GIS

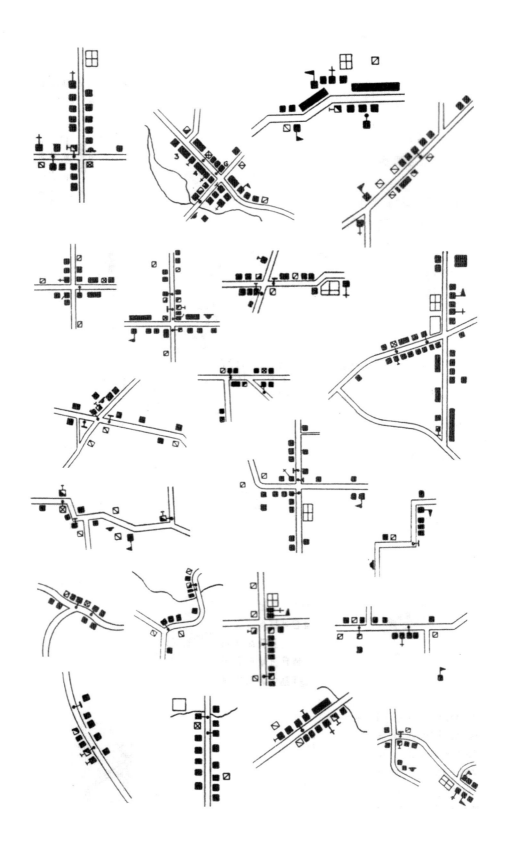

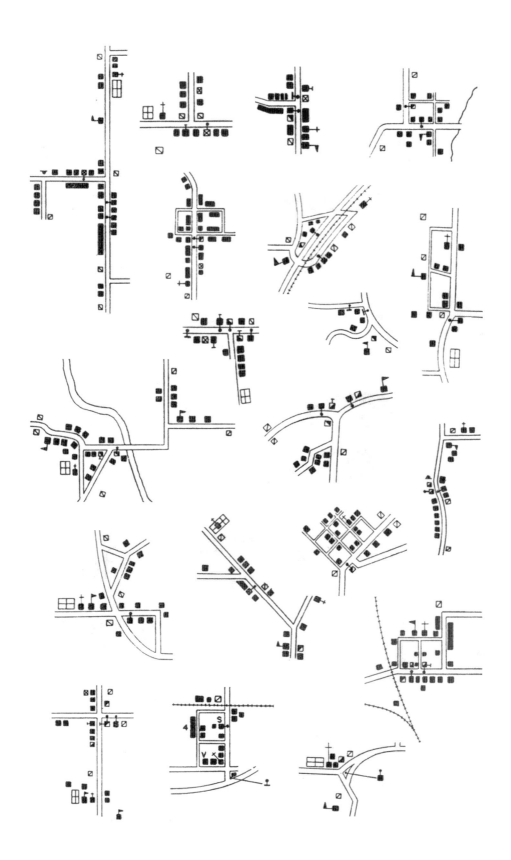

In December 1986 an experimental aircraft
named Voyager became the first piloted
aircraft to circle the earth without refueling.

	DAY 9				DAY 8				DAY 7				DAY 6				DAY 5		
Hours Aloft	216 hours	200	192 hours	184	176	168 hours	160	152	144 hours	136	128	120 hours	112	104	96 hours

Fuel on landing: 18 gallons

100° W 60° W 20° W 0° 20° E 60° E

United States

40° N

WNW

Triumphant landing
at Edwards AFB

Atlantic Ocean

20° N

Engine stalled;
unable to restart
for five harrowing
minutes

NNW
20

ENE
18

ESE
14

Oil warning
light goes on

Passing between
two mountains,
Rutan and Yeager
weep with relief
at having survived
Africa's storms

Worried about flying
through restricted
airspace, Rutan and
Yeager mistake the
morning star for
a hostile aircraft

Coolar
seal lea

Nicaragua

Rutan disabled
by exhaustion

Ethiopia

Somalia

W

Costa Rica

NW
10-15

Cameroon

E
10-20

0°

Transition
from tailwinds
to headwinds

E
34

Uganda

E
20

Gabon

Kenya

Squall line

Congo Zaire

Tanzania

E
37

Thunderstorm
forces Voyager
into 90° bank

Discovery
of backwards
fuel flow

20° S

Pacific Ocean

Flying among
'the redwoods':
life and death
struggle to avoid
towering
thunderstorms

Atlantic Ocean

40° S

120° W 80° W 40° W 0° 40° E

Visibility

Altitude
(feet)

20,000
15,000
10,000
5,000
sea level

Distance	26,678 miles traveled	5,000 miles to go	10,000 miles to go 12,532 miles previous record

Flight data courtesy of Len Snellman and Larry Burch, Voyager meteorologists
Mapped by David DiBiase and John Krygier, Department of Geography, University of Wisconsin-Madison, 1987

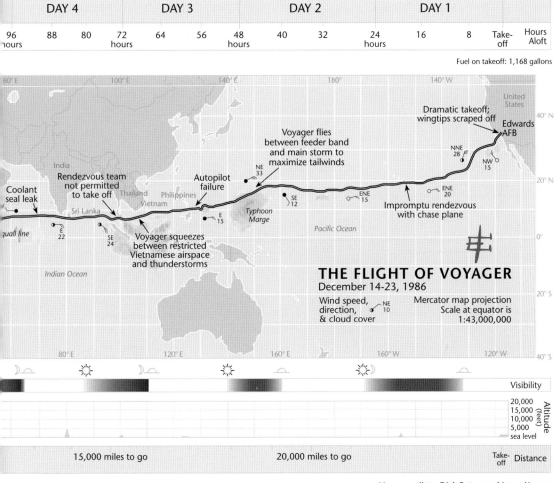

DAY 4				DAY 3				DAY 2				DAY 1			
96 hours	88	80	72 hours	64	56	48 hours	40	32	24 hours	16	8	Take-off	Hours Aloft

Fuel on takeoff: 1,168 gallons

60° E   100° E   140° E   180°   140° W

United States

Dramatic takeoff;
wingtips scraped off

Edwards
AFB

40° N

Voyager flies
between feeder band
and main storm to
maximize tailwinds

NNE
28

NW
15

NE
33

20° N

India

Rendezvous team
not permitted
to take off

Autopilot
failure

Coolant
seal leak

Thailand   Philippines

Vietnam

Sri Lanka

SE
12

ENE
15

ENE
20

Impromptu rendezvous
with chase plane

Typhoon
Marge

E
15

squall line

E
22

SE
24

Voyager squeezes
between restricted
Vietnamese airspace
and thunderstorms

Pacific Ocean

0°

Indian Ocean

THE FLIGHT OF VOYAGER
December 14-23, 1986

20° S

Wind speed,
direction,
& cloud cover

NE
10

Mercator map projection
Scale at equator is
1:43,000,000

80° E   120° E   160° E   160° W   120° W   40° S

Visibility

20,000
15,000
10,000
5,000
sea level

Altitude (feet)

15,000 miles to go   20,000 miles to go   Take-off   Distance

Voyager pilots: Dick Rutan and Jeana Yeager
Voyager designer: Burt Rutan

What do you need to know to make this map?

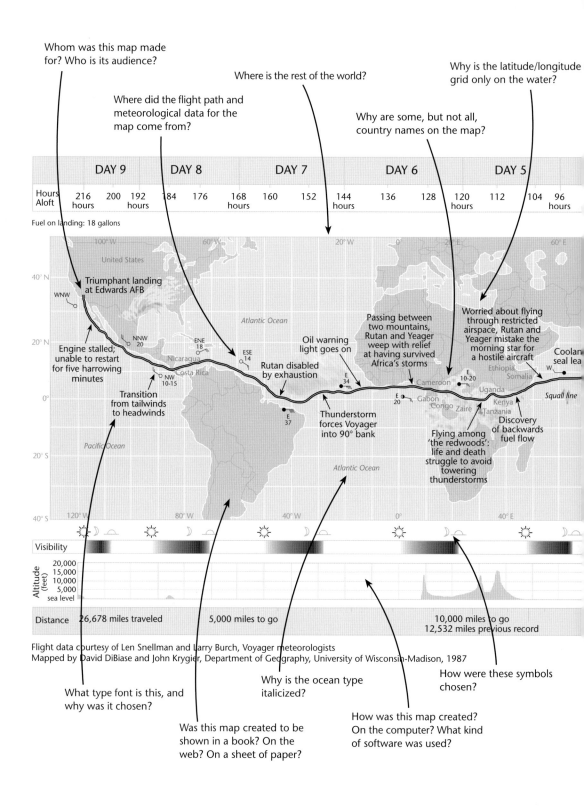

Whom was this map made for? Who is its audience?

Where did the flight path and meteorological data for the map come from?

Where is the rest of the world?

Why are some, but not all, country names on the map?

Why is the latitude/longitude grid only on the water?

	DAY 9			DAY 8			DAY 7		DAY 6			DAY 5		
Hours Aloft	216 hours	200	192 hours	184	176	168 hours	160	152	144 hours	136	128	120 hours	112	104	96 hours

Fuel on landing: 18 gallons

100° W   60° W   20° W   0°   20° E   60° E

United States

40° N

WNW

Triumphant landing at Edwards AFB

Atlantic Ocean

Engine stalled; unable to restart for five harrowing minutes

20° N

NNW 20

ENE 18

ESE 14

Oil warning light goes on

Passing between two mountains, Rutan and Yeager weep with relief at having survived Africa's storms

Worried about flying through restricted airspace, Rutan and Yeager mistake the morning star for a hostile aircraft

Coolan seal lea

Nicaragua

Rutan disabled by exhaustion

E 34

Ethiopia

E 10-20

W

Somalia

Costa Rica

NW 10-15

Cameroon

Transition from tailwinds to headwinds

0°

E 37

Thunderstorm forces Voyager into 90° bank

E 20

Gabon

Uganda

Kenya

Squall line

Congo Zaire

Tanzania

Flying among 'the redwoods': life and death struggle to avoid towering thunderstorms

Discovery of backwards fuel flow

Pacific Ocean

20° S

Atlantic Ocean

120° W   80° W   40° W   0°   40° E

40° S

Visibility

Altitude (feet): 20,000 15,000 10,000 5,000 sea level

Distance   26,678 miles traveled   5,000 miles to go   10,000 miles to go
12,532 miles previous record

Flight data courtesy of Len Snellman and Larry Burch, Voyager meteorologists
Mapped by David DiBiase and John Krygier, Department of Geography, University of Wisconsin-Madison, 1987

What type font is this, and why was it chosen?

Was this map created to be shown in a book? On the web? On a sheet of paper?

Why is the ocean type italicized?

How were these symbols chosen?

How was this map created? On the computer? What kind of software was used?

Why isn't there color on the map? Would color make the map better?

Where did data for the storms and typhoons come from?

Why are the days running backwards on the map?

Why is this line darker than other lines on the map?

Isn't every map supposed to have a north arrow?

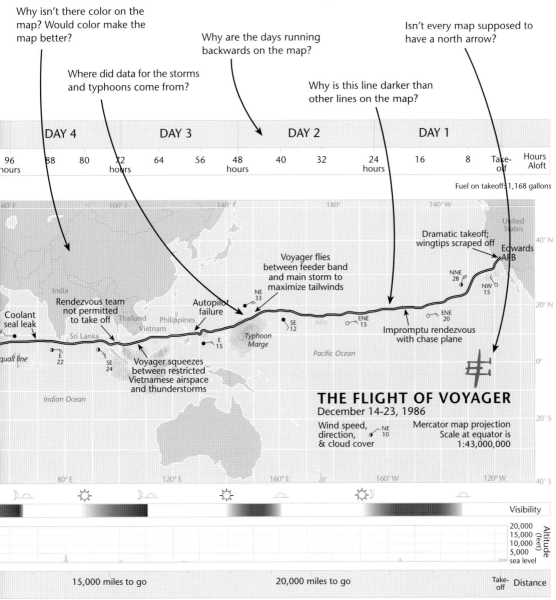

DAY 4				DAY 3				DAY 2			DAY 1			

| 96 hours | 88 | 80 | 72 hours | 64 | 56 | 48 hours | 40 | 32 | 24 hours | 16 | 8 | Take-off | Hours Aloft |

Fuel on takeoff: 1,168 gallons

United States

Dramatic takeoff; wingtips scraped off

Edwards AFB

40° N

NNE 28

NW 15

Voyager flies between feeder band and main storm to maximize tailwinds

NE 33

India

Rendezvous team not permitted to take off

Autopilot failure

20° N

Coolant seal leak

Thailand

Philippines

Vietnam

Sri Lanka

ENE 20

ENE 15

Impromptu rendezvous with chase plane

SE 12

SE 24

E 15

Typhoon Marge

Pacific Ocean

squall line

E 22

Voyager squeezes between restricted Vietnamese airspace and thunderstorms

Indian Ocean

0°

THE FLIGHT OF VOYAGER
December 14-23, 1986

Wind speed, direction, & cloud cover

NE 10

Mercator map projection
Scale at equator is
1:43,000,000

20° S

80° E 120° E 160° E 160° W 120° W 40° S

Visibility

20,000
15,000
10,000
5,000
sea level

Altitude (feet)

15,000 miles to go	20,000 miles to go	Take-off	Distance

Voyager pilots: Dick Rutan and Jeana Yeager
Voyager designer: Burt Rutan

CHAPTER

1

How to Make a Map

Start by looking; what do you see? Looking at maps is easy. Not really. You can glance at the Mona Lisa in a second. But to *get* the Mona Lisa you have to look more carefully. What do you see on the Voyager map? Words, lines, continents, a grid. A story, some information with the story. What do you notice first? Black lines, gray lines, white lines ... why are they different? Making maps requires that you answer such questions, and many more. Throughout this book, in nearly every chapter, we annotate *The Flight of Voyager.* By the end of the book, you will understand how to really see – and make – a map.

Making Maps is Hard

Whether looking at or making maps, there is a lot to see, think about, and do. Throughout this book, myriad subjects are considered in general and in relation to *The Flight of Voyager* map. A systematic critique of an existing map or the successful making of your own map is accomplished by considering the following issues. When making maps, think about everything before starting; then, when your map is complete, reconsider them all once again.

The Whole Map

Write out exactly what the map is supposed to accomplish: does the map meet its goals?

Are you sure a map is necessary?

Is the map suitable for the intended audience? Will the audience be confused, bored, interested, or informed?

Look at the map in its final medium: does it work? Has the potential of a black-and-white or color design been reached?

Is the map, its authors, its data, and any other relevant information documented and accessible to the map reader?

Look at the map and assess what you see; is it:

    confusing or clear
    interesting or boring
    lopsided or balanced
    amorphous or structured
    light or dark
    neat or sloppy
    fragmented or coherent
    constrained or lavish
    crude or elegant
    random or ordered
    modern or traditional
    hard or soft
    crowded or empty
    bold or timid
    tentative or finished
    free or bounded
    subtle or blatant
    flexible or rigid
    high or low contrast
    authoritative or unauthoratative
    complex or simple
    appropriate or inappropriate

Given the goals of the map, are any of these impressions inappropriate?

The Map's Data

Do the data serve the goals of the map?

Is the relationship between the data and the phenomena they are based on clear?

Does the map symbolization reflect the character of the phenomena or the character of the data?

Does the origin of the data – primary, secondary, tertiary – have any implications?

Are the data too generalized or too complex, given the map's goals ?

Is the map maker's interpretation of the data sound?

Are qualitative and quantitative characteristics of the data effectively symbolized?

Have the data been properly derived?

Has the temporal character of the data been properly understood and symbolized?

Is the scale of the map (and inset) adequate, given the goals of the map?

What about the accuracy of the data? Are the facts complete? Are things where they should be? Does detail vary? When were the data collected? Are they from a trustworthy source?

Have you consulted metadata (data about data)?

Does the map maker document copyright issues related to the data?

Is the map copyright or copyleft licensed?

The Map's Framework

What are the characteristics of the map's projection, and is it appropriate for the data and map goals? What is distorted?

Is the coordinate system appropriate and noted on the map?

The Design of the Map

Does the title indicate what, when, and where?

Is the scale of the map appropriate for the data and the map goals? Is the scale indicated?

Does textual explanation or discussion on the map enhance its effectiveness?

Does the legend include symbols that are not self-explanatory?

If the orientation of the map is not obvious, is a directional indicator included?

Are authorship and date of map indicated?

Are inset and locator maps appropriate?

Is the goal of the map promoted by its visual arrangement, engaging path, visual center, balance, symmetry, sight-lines, and the grid?

Has the map been thoroughly edited?

Does the map contain non data ink?

Has detail been added to clarify?

Do the data merit a map?

Do variations in design reflect variations in the data?

Is the context of the map and its data clear?

Are there additional variables of data that would clarify the goals of the map?

Do visual differences on the map reflect data differences?

Do important data stand out as figure, and the less important as ground, on the map? Are there consequences of data not included on the map?

Have visual difference, detail, edges, texture, layering, shape and size, closure, proximity, simplicity, direction, familiarity, and color been used to reflect figure-ground relationships appropriate to the map's goals?

Are the level of generalization and the data classification appropriate, given the map's goals?

Do map symbols work by resemblance, relationship, convention, difference, standardization, or unconvention? Are the choices optimal for the map's goals?

How do the map symbols relate to the concepts they stand for? Is the relationship meaningful?

Have the map symbols been chosen to reflect the guidelines suggested by the visual variables?

If symbolizing data aggregated in areas, is the most appropriate method used? How will the choice affect the interpretation of the map?

What do the words on your map mean? How do they shape the meaning of the map?

Has the chosen typeface (font) and its size, weight, and form effectively shaped the overall impression of the map as well as helping to symbolize variations in the data?

Does the arrangement of type on the map clarify, as much as possible, the data and the goals of the map?

Do color choice and variation reflect data choice and variation on the map?

Is color necessary for the map to be successful? Does color add anything besides decoration?

Do color choices grab viewer's attention while being appropriate for your data?

Does the map's design reflect the conditions under which it will be viewed?

Are color interactions and perceptual differences among your audience accounted for?

Have symbolic and cultural color conventions been taken into account and used to enhance the goals of the map?

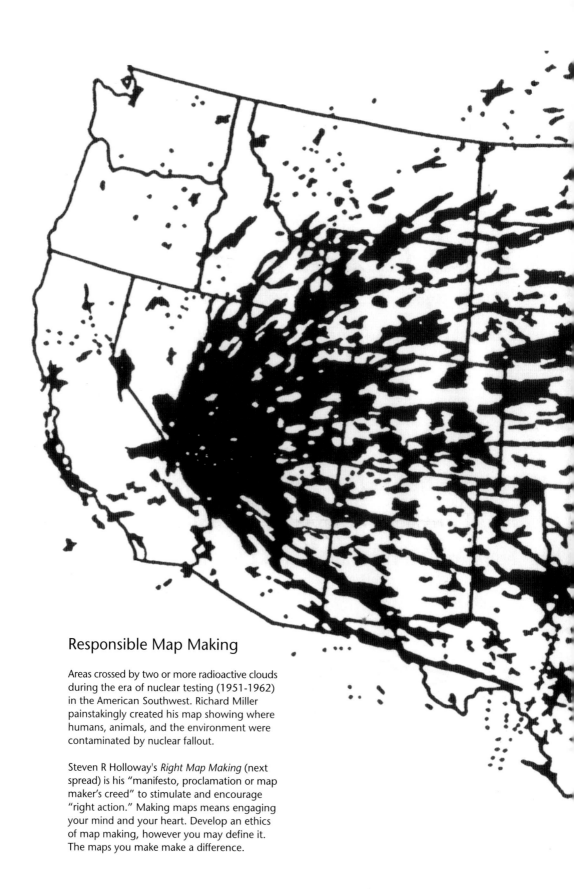

Responsible Map Making

Areas crossed by two or more radioactive clouds during the era of nuclear testing (1951-1962) in the American Southwest. Richard Miller painstakingly created his map showing where humans, animals, and the environment were contaminated by nuclear fallout.

Steven R Holloway's *Right Map Making* (next spread) is his "manifesto, proclamation or map maker's creed" to stimulate and encourage "right action." Making maps means engaging your mind and your heart. Develop an ethics of map making, however you may define it. The maps you make make a difference.

9

RIGHT MAP Making

"The most obvious characteristic of our age is its destructiveness." TH. MERTON

THE PROBLEM for the maker of maps being that our maps are, in part, engaged in the active and wanton destruction of the world.

Thus AWAKENED, we vow to take right effort & Engage in cartographic disobedience, map making "for a future to be possible." T. N. HANH Unacceptable it is not to ACT.

Five Ways to MAKE MAPS for a Future to be Possible

REVERENCE; the first precept of right map making

From the awareness that our maps are, in part, responsible for the great and unnecessary destruction of life taking place in the world today. We vow to map and comment on spatial relationships in a manner non-harming, with reverence and with respect, and to reflect and reveal the beauty of life in a manner non-objectified, where the economic, the non-economic, and the unseen elements are given voice. We vow to recognize and incorporate story with the arguments on our maps. In agreement with M. Gandhi, "first... non-cooperation with everything humiliating," we vow to refrain from economicism, the objectification of sentient beings, and cartographic pornography. Such mapping and maps reflect agreement with the first principle of right action: REVERENCE.

THE PRACTICE OF GENEROSITY; *the Second precept*

From the awareness that our maps are, too often, in our self-interest, greedy consumptions of endless desire, human biased and nationalistic. We vow to engage in a mapping of that which desires to be mapped and shared, not taking that into map form that which does not belong to us; desiring to remain unmapped. We vow to be generous to all sentient beings on our maps and in our mapping. Where generosity is also the courage to leave blank on the page that which does not belong to us, not mapping to take what is not ours, and honoring the sancity of the commons. Leviticus: *"fields are not to be reaped to the border."* Such mapping and maps show agreement with the second principle of right action: GENEROSITY.

COMMITMENT TO THE RELATIONSHIP WITH THE PLACE; *the third precept*

From the awareness that our maps are, in part, reflective of a lack of relationship and commitment to the place in which we reside and map. We vow to resist the temptation to map places with which we have no intimate or committed relation. We seek to remember and honor our relationship to the place; mapping with an honesty of lines, colours and shapes, the naming of places, the unnaming as well, without gossip or intent to harm, or to divide, but rather with a clarity of intent to all sentient beings with whom we are committed to with & in the relationship. Such mapping and maps show agreement with the third principle of right action: COMMITMENT TO THE RELATIONSHIP WITH THE PLACE.

DEEP LISTENING THROUGH DIRECT ~ CONTACT & STOPPING; *the fourth precept*

From the awareness that our maps are, in part, a failure to deeply listen and have been made without stopping to directly contact and listen to the place we are mapping. We vow to refrain from mapping what we do not know to be the truth, to first stop to experience the interconnected, ever-changing and interwoven space we are privileged to map. These maps acknowledge the intimate Other, the desire for the awakened heart and mind with & in direct contact with the place itSelf. Such mapping and maps show agreement with the fourth principle of right speech: DEEP LISTENING THROUGH DIRECT-CONTACT AND STOPPING.

ON BELONGING TO ONE BODY; *the fifth precept for a future to be possible*

From the awareness that our maps are, in part, disconnected from the body of the earth. How can this be? Kabir says, *"Whose Body is it anyway?"* We vow to make our maps about the body living; our own body, the body in motion, ever-changing and interconnected, the body free from addiction and enslavement to the toxicity of drugs: ownership, objectification, disconnection, greed, capitalism, all the *isms*. We vow to map that delight in the body that serves to reduce suffering and misery. Maps, and the making of maps that respect all sentient beings; the living breathing air, the changing clouds, and the wind and the tides in motion, the soils, the interwoven rocks, the waterways and the one Body without separation. Maps respecting and awakened to belonging to the OneBody with the fifth principle, oikos as the ecologic, economic and ecumenical whole of right livelihood: BELONGING TO ONE BODY.

Who died and made you the map police?

Jill, *Home Improvement* (1991)

For the execution of the voyage to the Indies, I did not make use of intelligence, mathematics or maps.

Christopher Columbus, *Book of Prophecies* (15th century)

I presume you have reference to a map I had in my room with some X's on it. I have no automobile. I have no means of conveyance. I have to walk from where I am going most of the time. I had my applications with the Texas Employment Commission. They furnished me names and addresses of places that had openings like I might fill, and neighborhood people had furnished me information on jobs I might get.... I was seeking a job, and I would put these markings on this map so I could plan my itinerary around with less walking. Each one of these X's represented a place where I went and interviewed for a job.... You can check each one of them out if you want to.... The X on the intersection of Elm and Houston is the location of the Texas School Book Depository. I did go there and interview for a job. In fact, I got the job there. That is all the map amounts to.

Lee Harvey Oswald, *Interrogation after Kennedy assassination* (November 24, 1963)

More...

The blog for this book, makingmaps.net, contains a curious collection of materials on maps and mapping and serves as an extension of this book. Check out cartotalk.com, a great discussion forum about maps and map design chock-full of cool map people.

Engage your *thinking* about maps: Jeremy Crampton, *Mapping: A Critical Introduction to Cartography and GIS* (Wiley-Blackwell, 2010); Brian Harley, *The New Nature of Maps* (Johns Hopkins University Press, 2002); Alan MacEachren, *How Maps Work* (Guilford Press, 2004); Mark Monmonier, *How to Lie with Maps* (University of Chicago Press, 1996); and Denis Wood, *Rethinking the Power of Maps* (Guilford Press, 2010). For a terrific overview of the diversity of maps througout history, see Brian Harley and David Woodward's multi-volume *History of Cartography* (1987-date, University of Chicago Press) series. Tony Campbell's website www.maphistory.info is a tremendous resource for the history of mapping.

This book, like all books, draws from numerous other texts, old and new, that can be consulted for more information than you'll ever want or need: R.W. Anson and F.J. Ormeling (eds.), *Basic Cartography* (International Cartographic Association, 1984); Borden Dent, Jeff Torguson, and Thomas Hodler, *Cartography: Thematic Map Design* (McGraw-Hill, 2008); J.S. Keates, *Cartographic Design and Production* (Wiley, 1973); Menno-Jan Kraak and F.J. Ormeling, *Cartography: Visualization of Spatial Data* (Longman, 1996); Juliana Muehrcke, A. Jon Kimerling, Aileen Buckley, and Phillip Muehrcke, *Map Use: Reading and Analysis* (ESRI Press, 2009); Arthur Robinson, Joel Morrison, Phillip Muehrcke, and A. Jon Kimerling, *Elements of Cartography* (Wiley, 1995); Erwin Raisz, *General Cartography* (McGraw-Hill, 1938) and *Principles of Cartography* (McGraw-Hill, 1962); Terry Slocum, Robert McMaster, Fritz Kessler, and Hugh Howard, *Thematic Cartography and Geovisualization* (Prentice Hall, 2008); and Judith Tyner, *Principles of Map Design* (Guilford Press, 2010). These folks *are* the "map police."

Check out the journal *Cartographic Perspectives* and the North American Cartographic Information Society (nacis.org), the journal *Cartographica* and the Canadian Cartographic Association (cca-acc.org), the *Cartographic Journal* and the British Cartographic Society (www.cartography.org.uk), and the International Cartographic Association (icaci.org).

Sources: Richard Miller, "Areas crossed by two or more radioactive clouds during the era of nuclear testing in the American Southwest, 1951-62" in *Under the Cloud: The Decades of Nuclear Testing* (Two-Sixty Press, 1999), between chapters 4 and 5. "Right MAP Making" copyright 2007 by Steven R Holloway. Designed and produced by toMake.com Press. "Right MAP Making" is intended to articulate the fundamental principles of ethical conduct in mapping and maps and to stimulate "right action." Set in Operina and Dante and printed from a freely distributed digital file. Forty letterpress copies are signed and numbered by the author. Editioned on the occasion of the 2007 Pecha Kucha of the North American Cartographic Information Society.

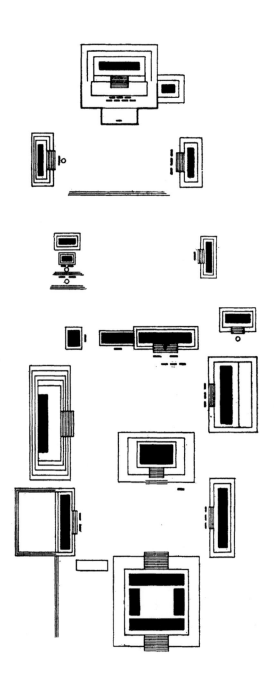

PODUNK

RIVER SEINE

FORT

VINCENNES

PLUNGE

What's the point?

The point, according to Mark Twain, is...

Inasmuch as this is the first time I ever tried to draft and engrave a map, or attempt anything in the line of art at all, the commendations the work has received and the admiration it has excited among the people have been very grateful to my feelings. And it is touching to reflect that by far the most enthusiastic of these praises have come from people who know nothing at all about art.

By an unimportant oversight I have engraved the map so that it reads wrong end first, except to left-handed people. I forgot that in order to make it right in print it should be drawn and engraved upside down. However, let the student, who desires to contemplate the map stand on his head or hold it before her looking-glass. That will bring it right.

The reader will comprehend at a glance that that piece of river with the "High Bridge" over it got left out to one side by reason of a slip of the graveing-tool which rendered it necessary to change the entire course of the River Rhine or else spoil the map. After having spent two days in digging and gouging at the map, I would have changed the course of the Atlantic Ocean before I would have lost so much work.

I never had so much trouble with anything in my life as I did with this map. I had heaps of little fortifications scattered all around Paris, at first, but every now and then my instruments would slip and fetch away whole miles of batteries, and leave the vicinity as clean as if the Prussians had been there.

The reader will find it well to frame this map for future reference, so that it may aid in extending popular intelligence and dispelling the widespread ignorance of the day.

MARK TWAIN.

OFFICIAL COMMENDATIONS.

It is the only map of the kind I ever saw.—*U. S. Grant.*

It places the situation in an entirely new light.—*Bismarck.*

I cannot look upon it without shedding tears.—*Brigham Young.*

My wife was for years afflicted with freckles, and though everything was done for her relief that could be done, all was in vain. But, sir, since her first glance at your map, they have entirely left her. She has nothing but convulsions now.—*J. Smith.*

It is very nice, large print.—*Napoleon.*

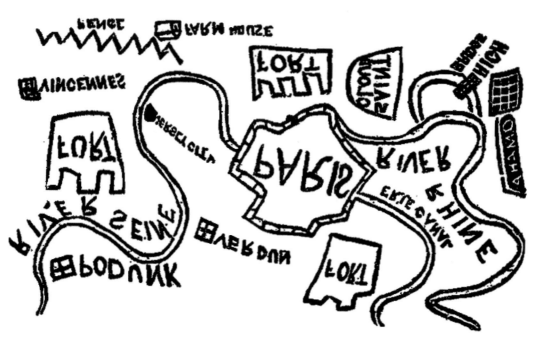

2 What's Your Map For?

What was Twain's map of Paris for? To make us laugh. But first it was to make Twain laugh. It was a dark time for Twain. "He swung between deep melancholy and half insane tempests and cyclones of humor." In one of the latter moments, "he got a board and with a jackknife carved a 'crude and absurd' map of Paris under siege." The map was a parody of those found in the newspapers of the time and was wildly popular. Who's your map for? How will you show it? How will you document, evaluate, and review it? Your answers will profoundly shape your map.

But Do You Really Need a Map?

The first thing you need to decide is whether you need a map. You may not. There are secrets that don't want to be mapped. There are circumstances where maps are inappropriate. And sometimes there are more effective ways of making your point: a graph, a drawing, a photo.

The Secret

Sometimes it's better not to map stuff you could easily map. Military sites, sacred indigenous locations, and archaeological sites are often left off of maps.

The U.S. Geological Survey topographic map of Raven Rock Mountain in Pennsylvania (above) doesn't show the extensive infrastructure of "Site R" – the bunker where U.S. Vice President Dick Cheney hunkered down after 9/11. Architect John Young tracked down the missing data and mapped it as part of his cryptome.org *Eyeballing* project (below).

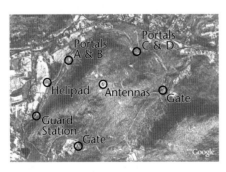

The Silly

"How surprised are you that Chicago has been eliminated from the potential host cities for the 2016 Olympics?"

■ Stunned
▫ Not surprised *ESPN SportsNation Poll & Map*

The Not Mappable

Typically land claims by native peoples are accompanied by maps. This is so obviously the place for a map that it seems perverse to question it, but increasingly Indigenous peoples have been arguing that maps can't capture their relationship to the land.

In 1987 the Gitxsan and the Wet'suwet'en in British Columbia entered the Gitxsan adaawk (a collection of sacred oral traditions about their ancestors, histories, and territories) and the Wet'suwet'en kungax (a spiritual song or dance or performance tying them to the land) as evidence in their suit seeking title to their ancestral lands. In 1997 the Canadian Supreme Court found that forms of evidence like these had to be accepted in Canadian courts.

Who's Your Map For?

Knowing the intended audience for your map will help you design it. Your audience may or may not be familiar with the area being mapped, an expert on the mapped topic or a novice, an eight-year-old or a college student. In each case, consider how your map can function better for the people who will actually use it.

Experts

Experts know a lot about the subject of the map. Experts are highly motivated and very interested in the facts the map presents. They expect more substance and expect to engage a complex map.

Less peripheral information on map explaining content and symbols

More information, more variables of information, more detail

Follow conventions of experts: consider using a spectral (rainbow) color scheme for ordered data if the user is accustomed to using such colors to show ordered data (such schemes are usually not good for other users)

Novices

Novices know less about the map subject and may not be familiar with the way maps are symbolized. They need a map that is more explanatory. Novices may be less motivated than expert users, but they want the map to help them learn something.

More peripheral information on map explaining content and symbols

Less information, fewer variables of information, less detail

Follow map design conventions, which enhance comprehension of the map

Mike...

```
                          ( 1 )( 1 )( 2 )( 2 )( 3 )( 3 )
                          ( 1 )( 1 )( 2 )( 2 )( 3 )( 3 )
                            ( 4 )( 4 )( 5 )( 5 )( 6 )( 6 )
                            ( 4 )( 4 )( 5 )( 5 )( 6 )( 6 )
                          ( 7 )( 7 )( 8 )( 8 )( 9 )( 9 )( 10 )( 10 )( 11 )( 11 )
                          ( 7 )( 7 )( 8 )( 8 )( 9 )( 9 )( 10 )( 10 )( 11 )( 11 )
                        ( 12 )( 13 )( 14 )( 15 )( 16 )( 17 )( 17 )( 18 )( 18 )( 19 )( 19 )
         (205)(205)        ( 20 )( 21 )( 22 )( 23 )( 24 )( 25 )( 17 )( 17 )( 18 )( 18 )( 19 )( 19 )
         (205)(205)( 26 )( 27 )( 28 )( 29 )( 30 )( 31 )( 32 )( 33 )( 34 )( 35 )( 36 )( 36 )( 37 )( 37 )( 38 )( 38 )
         (204)(204)( 39 )( 40 )( 41 )( 42 )( 43 )( 44 )( 45 )( 46 )( 47 )( 58 )( 36 )( 36 )( 37 )( 37 )( 38 )( 38 )
         (204)(204)( 49 )( 50 )( 51 )( 52 )( 53 )( 54 )( 55 )( 56 )( 57 )( 58 )( 59 )( 59 )( 60 )( 60 )( 61 )( 61 )( 62 )( 62)
( 63 )( 63 )( 64 )( 64 )( 65 )( 66 )( 67 )( 68 )( 69 )( 70 )( 71 )( 72 )( 73 )( 74 )( 59 )( 59 )( 60 )( 60 )( 61 )( 61 )( 62 )( 62 )
( 63 )( 63 )( 64 )( 64 )( 75 )( 76 )( 77 )( 78 )( 79 )( 80 )( 81 )( 82 )( 83 )( 84 )( 85 )( 85 )( 86 )( 86 )( 87 )( 87 )( 88 )( 88 )( 89 )( 89 )
( 90 )( 90 )( 91 )( 91 )( 92 )( 93 )( 94 )( 95 )( 96 )( 97 )( 98 )( 99 )(100)(101)( 85 )( 85 )( 86 )( 86 )( 87 )( 87 )( 88 )( 88 )( 89 )( 89 )
( 90 )( 90 )( 91 )( 91 )(102)(103)(104)(105)(106)(107)(108)(109)(110)(111)(112)(112)(113)(113)(114)(114)(115)(115)
(116)(116)(117)(117)(118)(119)(120)(121)(122)(123)(124)(125)(126)(127)(112)(112)(113)(113)(114)(114)(115)(115)
(116)(116)(117)(117)(128)(129)(130)(131)(132)(133)(134)(135)(136)(137)(138)(138)(139)(139)(140)(140)
(141)(141)(142)(142)(143)(144)(145)(146)(147)(148)(149)(150)(151)(152)(138)(138)(139)(139)(140)(140)
(141)(141)(142)(142)(153)(154)(155)(156)(157)(158)(159)(160)(161)(162)(163)(163)
(164)(164)(165)(165)(166)(167)(168)(169)(170)(171)(172)(173)(174)(175)(163)(163)
(164)(164)(165)(165)(176)(177)(178)(179)(180)(181)(182)(183)(184)(185)
         (186)(186)(187)(187)(188)(188)(189)(189)(190)(190)(191)(191)
         (186)(186)(187)(187)(188)(188)(189)(189)(190)(190)(191)(191)
         (192)(192)(193)(193)(194)(194)(195)(195)(196)(196)
         (192)(192)(193)(193)(194)(194)(195)(195)(196)(196)
         (197)(197)(198)(198)(199)(199)(200)(200)
         (197)(197)(198)(198)(199)(199)(200)(200)
            (201)(201)(202)(202)(203)(203)
            (201)(201)(202)(202)(203)(203)
```

Social worker Mike Rakouskas's map of Wake County, North Carolina. The numbers refer to pages in the county street atlas he uses, and the shaded numbers are client sites. He uses this map to rationalize his trip planning and as an index to the atlas. It was made with a word processor. Peculiar! Clever! And perfect for Mike.

How Are You Going to Show It?

Consider the final medium of your map before making it. Most maps are made on computer monitors, but the monitor is not the final medium. Rather, it might be a cell phone screen, a piece of paper, a poster, a slide projected on a screen during a presentation, a yard sign, handbill, or protest sign. What looks great on your computer will probably not look so great when printed or projected or shown on a tiny phone screen.

A yard sign complete with map in opposition to a road that would really mess up a lot of stuff in Raleigh, North Carolina.

Black and White, on Paper

Most maps are created on computer monitors, with less resolution and area than is possible on a piece of paper. When paper is your final medium, design for the paper and not for the monitor. Always check design decisions by printing the map (or having your printer create a proof if your map is to be professionally printed). While all computers offer color, final printing with color is not always an option. Don't despair! Much can be done with black and white.

Map size should match final paper size, with appropriate margins

10-point type works well on a printed map, but you may have to zoom in to see it on the computer monitor

Point and line symbols can be smaller and finer on a printed map than on the computer

More subtle patterns can be used than on a computer monitor map

More data and more complex data can be included on a printed map

Substitute a range of grays and black and white for color. Remember that printers cannot always display as many grays as you can create on a monitor; subtle variations in grays may not print clearly

Black will be more intense than white; use white to designate no information or the background, dark to designate more important information

Monochrome copiers sometimes reproduce gray tones poorly

Very light gray tones may not print

Color, on Paper

Color on a computer monitor is created in a different manner than color on desktop printers or on professionally printed maps. Select colors on the computer, then print and evaluate (or ask for a proof). Always design for the final medium: adjust the colors on the monitor so they look best for the final output. The same colors will vary from printer to printer. Reproducing color is often more expensive than black and white. Finally, keep in mind that users may reproduce your color map in black and white. Will it still work?

Map size should match final paper size, with appropriate margins

10-point type works well on a printed map, but you may have to zoom to see it on the computer monitor

Point and line symbols can be smaller and finer on a printed map

More subtle patterns can be used than on a computer monitor map

More data and more complex data can be included on a printed map

Use color value (e.g., light red vs. dark red) to show differences in amount or importance. Use color hue (blue vs. red) to show differences in kind. Desktop printers cannot display as many colors as you can create on a monitor; subtle variations in colors may not print

Dark colors are more intense than light; use light colors to designate less important information and background, and dark to designate more important information

Never print a color map in black and white; redesign it for black and white

Computer Monitors

Designing maps for final display on a computer must take into account screen resolution and space limits. Desk or laptop computer monitor resolution is typically 72 dots per inch (dpi), compared to 1200 or more for many printers. Computer monitors also have limited area, typically 7 by 9 inches (gray area on this spread), or less if the map is displayed in a web browser window. Design a map so that all type and symbols are visible without magnification. Also avoid maps that require the viewer to scroll around to see the entire map. Use more than one map if you need more detail, or consider web tools that allow you to zoom and pan over a map.

The entire map should fit on the screen without scrolling (if pan/zoom is not possible)

Increase type size: 14 point type is the smallest you should use on a monitor

Make point and line symbols 15% larger than those on a paper map

Use more distinct patterns: avoid pattern variations that are too fine or detailed

You may have to limit the amount and complexity of data on your map, compared to a print map

Use color: but remember that some monitors cannot display billions of colors; subtle color variations may not be visible on every monitor

White is more intense than black. Take care when using white to designate the lack of information or as background color, it may stand out too much.

Save static maps for the internet at 72-150 dpi. Size the map to fit in a browser window

Design your map so it works on different monitors (RGB, LCD, portables)

Interactive maps require attention to additional issues such as pan, zoom, interactivity, etc.

Sony Ericsson Xperia

LG + Samsung

Centro

Treo

QVGA Type 2.2

Portable Monitors

Maps on smart phones, PDAs, GPS units, and other portable devices pose the same design challenges as on desktop monitors, with the further limitation of screen size. Typical portable monitor sizes are shown on this spread. Many portable monitors are touch-sensitive, allowing users to pan and zoom, thus overcoming some of the limitations of the small monitor size.

Static maps on portable devices can follow desktop monitor design guidelines, taking into account the limited display size

Interactive maps should use appropriate interface metaphors: zoom in is "up" on a slider bar, or two fingers diverging outward. Pan is touch and slide in the appropriate direction

Interactive maps should vary map design specifications with scale

Generalize more as the user zooms out on the map: for example, local roads and road names disappear when zoomed out

Generalize less as the user zooms in on the map: local roads and their names appear when zoomed in

Aerial photographs may be more appropriate than maps for users with limited navigation abilities

Maps may be more appropriate than aerial photographs for users with better navigation abilities

Ground-view images may be more helpful for navigation than maps alone, but using both should increase navigation success

Map symbols should not be too complex

Colors should be more intense to account for varying lighting conditions

Serif fonts may be easier to read on portable monitors than sans serif

25

Projections

It is increasingly common for maps to be shown on a large screen with a computer projector. When projected, white and lighter colors will be more intense, black and darker colors subdued. Computer projectors vary in the amount of light they can project. Some projectors wash out colors. Consider previewing your projected map and adjusting the projector. Projected maps must be designed with the viewing distance in mind (find out the size of the room). A map projected to an audience in a small room can have smaller type and symbols than a map projected in an auditorium. Always check that the map is legible from the back of the room in which the map will be displayed.

Greater map size is offset by the increased viewing distance

Increase type size so that smallest type is legible from the back of the room

Increase point and line symbol size to be legible from the back of the room

More distinct patterns: avoid pattern variations that are too fine or detailed

You may have to limit the amount and complexity of data on your map compared to print maps

Older or lower-output projectors may wash-out colors, so intensify your colors for projection

If your map will be projected in a dark room, use black as background, darker colors for less important information, and lighter colors for more important information

If your map will be projected in a well-lighted room, use white as background, lighter colors for less important information, and darker colors for more important information

Posters

Posters are similar to projected maps, although usually viewed in well-lighted conditions. Viewers should be able to see key components of the map (such as the title) from afar, then walk up to the map and get more detail. Design the poster, then, so information can be seen both close and at a distance. The size of poster maps is limited by the largest printer you can use; always check color and resolution of the printer used to reproduce your poster. You may want to request a test print of the colors you plan to use to evaluate your color choices.

Design map title and mapped area so they are legible from across the room

The majority of type, point, and line symbols should be slightly larger than on a typical printed map, but not as large as on a monitor or projected map. Design this part of the map so it is legible from an arm's length

More complex information can be included on a poster map than on a computer monitor or projected map

Follow color conventions for color printed maps

Most posters are viewed in a well-lighted room, so use white as background, lighter colors for less important information, and darker colors for more important information

Right: a portion of the poster-sized map *Guide Psychogeographique OWU*, made by a group of middle-school kids and John Krygier during a summer class, *Mapping Weird Stuff,* at Ohio Wesleyan University, June 2009.

Guide Psychogéographique de OWU

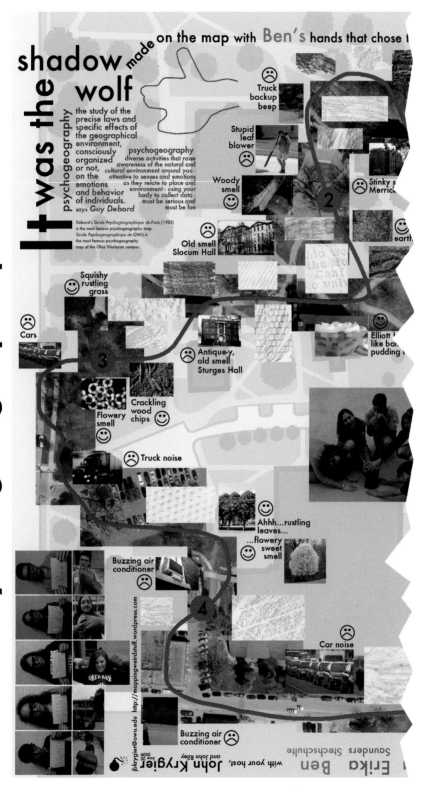

Document, Evaluate, Review

Constantly cast a critical eye on your work. Document what you do and continually evaluate whether the map is serving its intended goal, meeting the needs of its intended audience, and working well in its final medium.

NASA's Bob Craddock set about revising a 1986 map of Mars with new imagery from the Viking Orbiter. Craddock transferred details from the 1986 map while referencing his new data, drawing lines and labeling what he thought he saw, evaluating the data as he worked. Craddock used the old interpretations when the new data supported them, and modified features clarified by the new data.

When complete, the new map was sent to other experts for review and evaluation. The reviewers annotated the map wherever they disagreed with Craddock's interpretations or saw alternatives. Craddock, in turn, revised his map with the reviewers' comments, not necessarily agreeing with all of them but, in the end, producing a map of the geology of Mars that was better because of the expert evaluation and review.

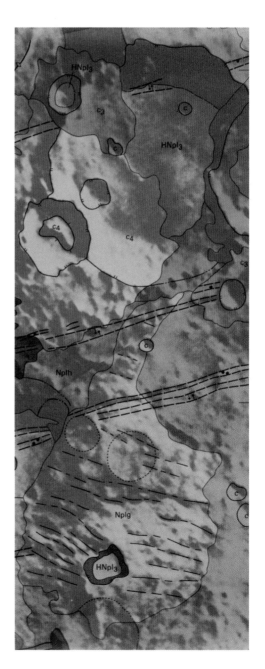

Documentation

What were those six great shades of red I used on that map I made last month? What font did I use on the last poster map I made? How big was the title type? How long did it take me to make that map for the annual report last year? Where did we get that great data set? Was it licensed? Who printed that large format map for us last year? How much did it cost to print and fold those color maps?

Documentation of the details involved in making a map may seem tedious but can save time and effort in future map making, both for yourself and others who may need to make similar maps. Working toward a few general styles that are effective for specific types of commonly produced maps is useful. Documentation of mapped data is vital if the map is to be published.

Documenting General Issues

Document your goal for the map and
...the intended audience, and what you know about them
...the final medium, and details about the medium that will affect map design and reproduction
...the amount of time it takes to create the map, and any major problems and how you solved them
Keep copies of the map as well as information on where it was published or presented

Documenting Data

Document the source of the data, including contact information and copyright information
...the age, quality, and any limitations of the data
...how the data were processed into a form appropriate for mapping
...map projection and coordinate system information

Documenting Design

Document specifics of map size, scale, and sketches of layouts
...a list of information on the map, arranged in terms of importance, and associated symbols
...data classification and generalization information
...sources and details of map symbols
...details of type size, font, etc.
...color specifications for all colors used
...design problems encountered and solutions
...software problems encountered and solutions

Formative Evaluation

Ongoing formative evaluation is as simple as asking yourself whether the map is achieving its goals throughout the process of making the map. Formative evaluation implies that you will "re-form" the map so it works better, or maybe even dump it! It is never too late to bail if the map is not serving your needs. It is a good idea to ask others to evaluate your map as well: What do you think of those colors? Can you read that type from the back of the room? Does what is most important on the map actually stand out? What is the boss going to think? Simply engaging your mind as you make your map, and being open to critique and change, will lead to a better map.

Ask yourself...

Is this map doing what I want it to do?

Will this map make sense to the audience I envision for it?

How does the map look when printed, projected, or viewed in the final medium, and what changes will make it better?

Are the chosen scale, coordinate system, and map projection appropriate?

Do the layout of the map and the map legend look good? Could it be adjusted to help make the map look better and easier to interpret?

Does the most important information on the map stand out visually? Does less important information fall into the background?

Are data on the map too generalized or too detailed, given the intent of the map?

Does the way I classified my facts help to make sense out of them? Would a different classification change the patterns much?

Do chosen symbols make sense, and are they legible?

Is the type appropriate, legible, and is its size appropriate, given the final medium?

Is color use logical (e.g., value for ordered data, hue for qualitative data) and appropriate, and will the chosen colors work in the final medium?

Do I want a series of simpler maps, or one more complicated one?

Is a handout map needed, if presenting a map on a poster or projected?

...then re-*form* your map.

Impact Evaluation

Impact evaluation is a range of informal and formal methods for evaluating the finished map. It may be your boss or a publisher reviewing the map, or public feedback on the map's efficacy. You should begin any map making with a clear sense of who may have the final say on the acceptability of your map, and factor in their wants, needs, and requirements at the beginning of the process.

Caribou Calving Areas
Arctic National Wildlife Refuge (ANWR)

Percent Likelihood of High Density Calving Area

- 50% - 75%
- 25% - 49%
- 1% - 24%

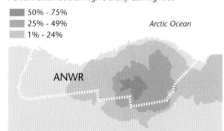

Arctic Ocean

ANWR

Ian Thomas, a contractor for the U.S. Geological Survey, was fired, allegedly for making maps of caribou calving areas in the ecologically and politically sensitive Arctic National Wildlife Refuge. Thomas argues he was fired for publicizing facts that would undermine the push for oil exploration in the refuge. Others claim the maps were of out-of-date information beyond Thomas's area of expertise and had nothing to do with his firing. In either case, it is obvious that making maps can piss off your boss.

An old Japanese map from the David Rumsey digital map collection was added to Google Earth in early 2009. A label on the map described a village as populated by "eta," the untouchable caste of burakumin (translation, "filthy mass"). Because some idiots in Japan discriminate against the burakumin, it is common practice to remove such references. Rumsey initially decided not to censor the map, but after an uproar the offending nomenclature was removed.

The Flight of Voyager map was published in 1987 in the book *Voyager* by Jeana Yeager, Dick Rutan, and Phil Patton

	DAY 9			DAY 8			DAY 7			DAY 6			DAY 5	

Hours Aloft	216 hours	200	192 hours	184	176	168 hours	160	152	144 hours	136	128	120 hours	112	104	96 hours

Fuel on landing: 18 gallons

Triumphant landing at Edwards AFB

Engine stalled; unable to restart for five harrowing minutes

Transition from tailwinds to headwinds

Oil warning light goes on

Rutan disabled by exhaustion

Thunderstorm forces Voyager into 90° bank

Passing between two mountains, Rutan and Yeager weep with relief at having survived Africa's storms

Flying among 'the redwoods': life and death struggle to avoid towering thunderstorms

Worried about flying through restricted airspace, Rutan and Yeager mistake the morning star for a hostile aircraft

Discovery of backwards fuel flow

United States · Atlantic Ocean · Nicaragua · Costa Rica · Pacific Ocean · Atlantic Ocean · Cameroon · Gabon · Congo · Zaire · Uganda · Kenya · Tanzania · Ethiopia · Somalia · Squall line · Coolan seal lea

Visibility

Altitude (feet): 20,000 / 15,000 / 10,000 / 5,000 / sea level

| Distance | 26,678 miles traveled | 5,000 miles to go | 10,000 miles to go
12,532 miles previous record |
|---|---|---|---|

Flight data courtesy of Len Snellman and Larry Burch, Voyager meteorologists
Mapped by David DiBiase and John Krygier, Department of Geography, University of Wisconsin-Madison, 1987

David DiBiase and John Krygier designed and made a map to tell the story of Voyager and its pilots. The map was created for a map design course at the University of Wisconsin-Madison taught by David Woodward.

The map was made for readers of the book *Voyager* (1987), with its general, educated audience, including those with a specialist interest in flight and aerospace. Given the audience, the map was designed to contain a significant amount of information including detailed data, sure to resonate with pilots.

The map was split between the front and back book end-papers, half in the front and half in the back. Each endpaper was 9" high and 12" wide.

The map was designed to be viewed at arms length.

DAY 4				DAY 3			DAY 2			DAY 1			
96 hours	88	80	72 hours	64	56	48 hours	40	32	24 hours	16	8	Take-off	Hours Aloft

Fuel on takeoff: 1,168 gallons

THE FLIGHT OF VOYAGER
December 14-23, 1986

Wind speed, direction, & cloud cover

Mercator map projection
Scale at equator is 1:43,000,000

Coolant seal leak

Rendezvous team not permitted to take off

Voyager squeezes between restricted Vietnamese airspace and thunderstorms

Autopilot failure

Voyager flies between feeder band and main storm to maximize tailwinds

Dramatic takeoff; wingtips scraped off

Impromptu rendezvous with chase plane

Typhoon Marge

squall line

India

Sri Lanka

Thailand

Vietnam

Philippines

Pacific Ocean

Indian Ocean

United States

Edwards AFB

Visibility

Altitude (feet)
20,000
15,000
10,000
5,000
sea level

15,000 miles to go

20,000 miles to go

Take-off Distance

Voyager pilots: Dick Rutan and Jeana Yeager
Voyager designer: Burt Rutan

The publisher of *Voyager*, Knopf, allowed us black and one color for the map. We chose deep red for the most important information (such as the flight path and related text). The map was redesigned in monochrome for *Making Maps, 2nd Ed.* The map still works!

Details of the map's design – line weights, type size, percent gray of different areas on the map, etc. – were documented, as we were taught in David Woodward's course and at the University of Wisconsin-Madison Cartographic Lab. Formative evaluation was ongoing throughout the process, and the editors at Knopf provided the final edit and evaluation of the map.

Maps are full of the impertinence of the arbitrary.

Brian O'Doherty, *American Masters* (1973)

The most remarkable escape story of all concerns Havildar Manbahadur Rai.... He escaped from a Japanese prison camp in southern Burma and in five months walked 600 miles until at last he reached the safety of his own lines. Interrogated by British intelligence about his remarkable feat, Manbahadur told them that ... he had a map, which before his capture had been given to him by a British soldier in exchange for his cap badge. He produced the much creased and soiled map. The intelligence officers stared at it in awe. It was a street map of London.

Byron Farwell, *The Gurkhas* (1984)

Borders are scratched across the hearts of men
By strangers with a calm, judicial pen,
And when the borders bleed we watch with dread
The lines of ink across the map turn red.

Marya Mannes, *Subverse: Rhymes for Our Times* (1959)

More...

If you're making your map for any of the usual reasons, there's more than enough to read in the books we listed at the end of Chapter 1. If you've got something else in mind, they're unlikely to be much help.

But there are a lot of other things to look at. Nato Thompson's *Experimental Geography: Radical Approaches to Landscape, Cartography, and Urbanism* (Melville House, 2008) is crammed with examples of novel maps and map-fusions intended to blow your complacency if not your mind. Lize Mogel and Alexis Bhagat's *An Atlas of Radical Cartography* (Journal of Aesthetics and Protest Press, 2007) is a book and portfolio of maps in which the authors "wear their politics on their sleeves." Denis Wood's *Everything Sings: Maps for a Narrative Atlas* (Siglio Press, 2010) is a sequence of maps intended to be read as poems or as chapters in a story. His introduction describes the difficulty he had in breaking out of the map mold. Frank Jacobs's *Strange Maps: An Atlas of Cartographic Curiosities* (Viking, 2009) is a collection less strange than the title implies, but nonetheless rich with suggestive directions.

Community and land use occupancy mapping is a whole other kettle of salmagundi. Two great textbooks are Terry Tobias's *Chief Kerry's Moose: A Guidebook to Land Use and Occupancy Mapping, Research Design, and Data Collection* (Union of BC Indian Chiefs and Ecotrust Canada, 2000) and Alix Flavelle's *Mapping Our Land: A Guide to Making Maps of Our Own Communities and Traditional Lands* (Lone Pine Foundation, 2002). In the U.K., Common Ground's Parish Maps Project reaches a similar destination along a wholly different path. Common Ground (www.commonground.org.uk) offers Sue Clifford and Angela King's useful pamphlet *From Place to PLACE: Maps and Parish Maps* (Common Ground, 1996), but Kim Leslie's extraordinary atlas, *A Sense of Place: West Sussex Parish Maps* (West Sussex County Council, 2006), not only has gloriously reproduced examples of nearly a hundred parish maps made for the millennium observance, but also many descriptions by residents of how they made them.

If you're interested in making maps as art, the place to start is with Katherine Harmon's *The Map as Art: Contemporary Artists Explore Cartography* (Princeton Architectural Press, 2009).

Sources: The Mark Twain map and text were originally published in the *Buffalo Express* newspaper on September 17, 1870. The *NO to Century Boulevard* yard sign is used courtesy of Ron Rozzelle. Mobile screen sizes from the *Sender11* blog. Bob Craddock's Mars map, created by him and Ronald Greeley, is titled "Geologic Map of the MTM -20147 Quadrangle, Mangala Valles Region of Mars" (U.S. Department of the Interior, U.S. Geological Survey, prepared for the National Aeronautics and Space Administration). The caribou calving map is re-created from data and maps at Ian Thomas's web site. The Google eta/burakumin map is used courtesy of Google and David Rumsey.

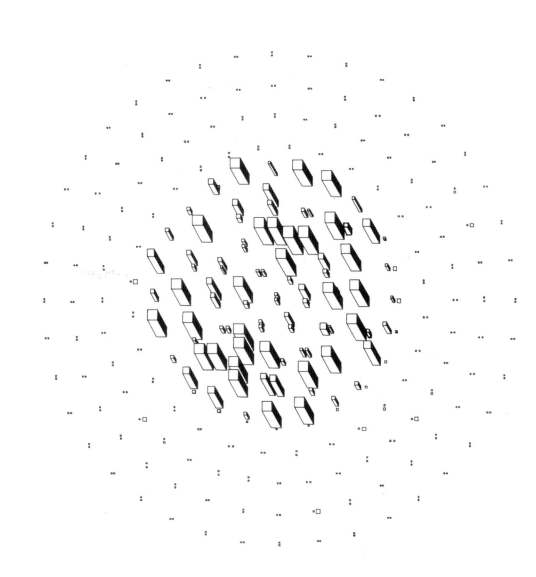

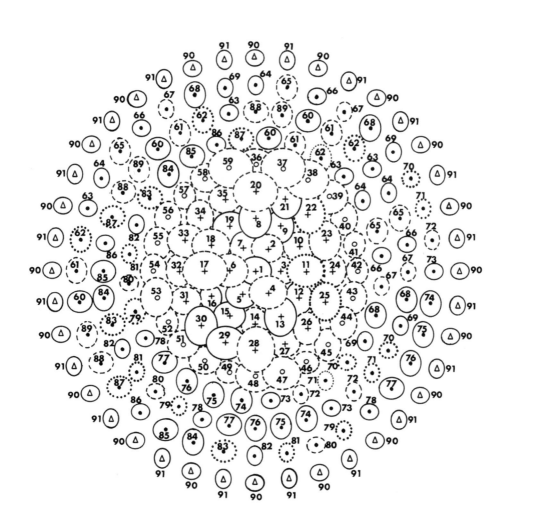

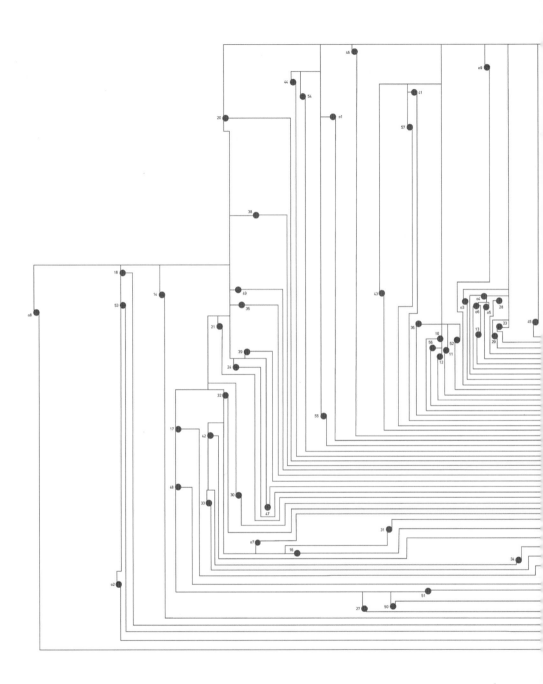

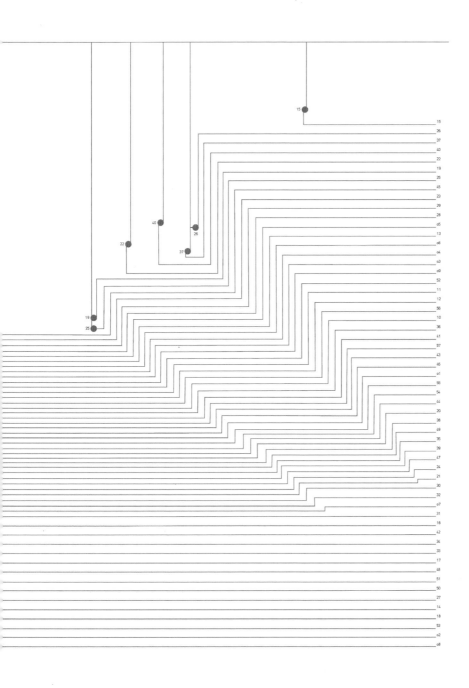

What do maps show us?

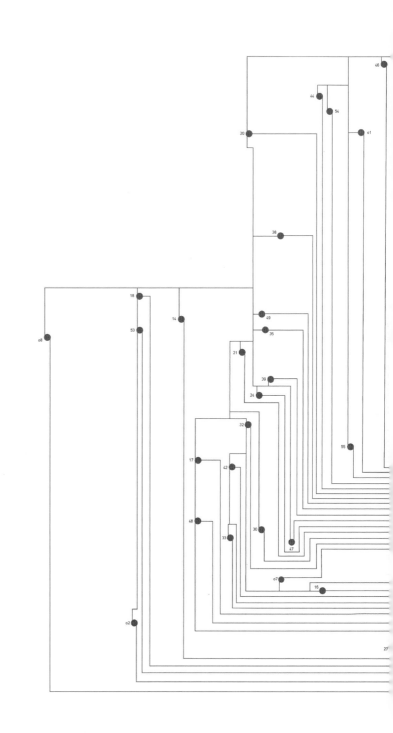

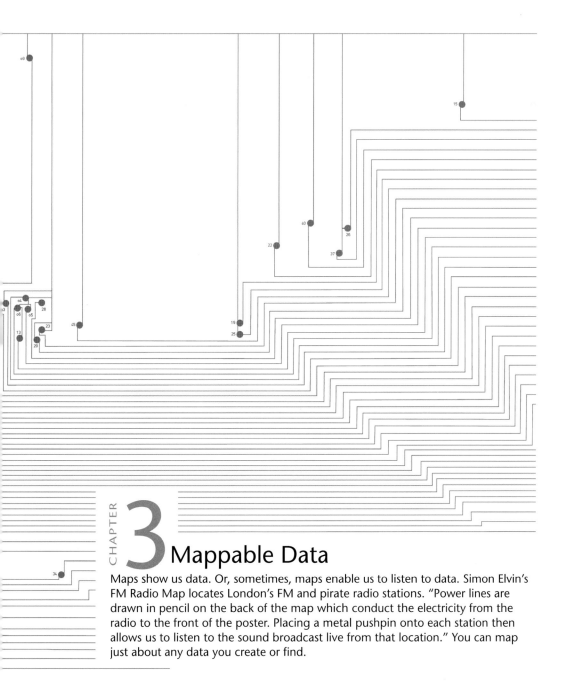

3 Mappable Data

Maps show us data. Or, sometimes, maps enable us to listen to data. Simon Elvin's FM Radio Map locates London's FM and pirate radio stations. "Power lines are drawn in pencil on the back of the map which conduct the electricity from the radio to the front of the poster. Placing a metal pushpin onto each station then allows us to listen to the sound broadcast live from that location." You can map just about any data you create or find.

Individual and Aggregate

Phenomena are all the stuff in the real world. Data are records of observations of phenomena. Maps show us data, not phenomena. Carefully consider the data you are mapping, how it relates to the stuff in the world it stands for, how it's similar, how it's different, and how that may affect our understanding of the phenomena. Differentiate, for example, between individual and aggregated data:

A map of individual trees smacked flat by a tornado in a forest. The phenomena are trees, and the data – individually mapped by orienteers – retain this individuality. The cross bar of the "t" point symbol is the root ball of the tree, indicating the direction the tree fell.

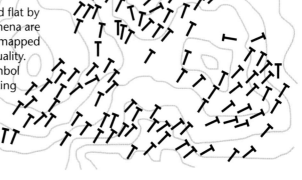

An old map shows the range of the pin oak tree in the U.S. The phenomena are not trees but the range of a species, and the data – aggregated from other maps – embody this abstraction.

A map of major vegetation zones in the eastern U.S. The area labeled "D" is broadleaf deciduous forest. The adjacent "M" is mixed broadleaf and needleleaf forest. The phenomena here are neither trees nor species, but forest, and the data – aggregated from many different maps – embody this still greater abstraction.

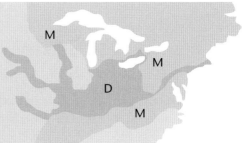

Aggregating data *changes phenomena*, as here, from individual trees, to species, to forest.

Continuous and Discrete

Here the issue is change, how the phenomena and/or data vary over space and time. Continuous phenomena vary gradually, continuously, more or less smoothly. Good examples are atmospheric pressure and temperature. Discrete phenomena change abruptly, like laws from one jurisdiction to another. There is no necessary relationship between phenomena and data, as it's possible to have many kinds of data for any given phenomenon.

We know that air temperature varies continuously, but thermometers can only record it at points. We can map it this way, revealing the structure of the data...

Sometimes we have continuous or nearly continuous data for more or less continuous phenomena but choose to map the data as discrete units. For convenience we map the continuous change in time around the globe into discrete time zones.

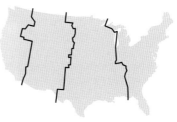

...or we can transform the point data into continuous data by interpolating data that may exist between the readings, revealing the structure of the phenomena.

For administrative reasons we usually map detailed census data into large single-value areas, often political.

There are many differences between real-world phenomena and the way they are portrayed on maps. Sometimes we believe it possible to intuitively show aspects of the phenomena, as here with eddies, kelp-beds, and half-sunken shipwrecks. At other times we simply display the data, as with a small circle indicating a snag; here words can do a lot of the work.

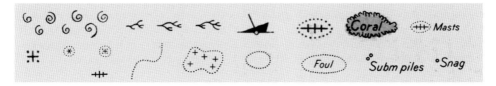

Creating and Getting Data

Data are records of observations of phenomena. These records may be made by machines (like those made by a recording thermometer) or by the map makers themselves. All these are primary data, that is, records of observations made in the environment itself. Maps made from primary data can be considered evidence. Most map makers use secondary or tertiary data sources created and published by others, but it's surprisingly easy to create mappable data yourself. It is common, and often necessary, to combine primary, secondary, and tertiary data sources on a map.

Primary Data Sources

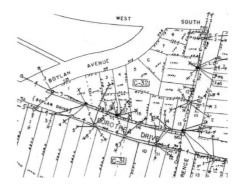

Collecting data at addresses. Researchers recorded 20 different facts about the exterior of houses (above: data collection sheet). Such data can be geocoded (address matched) using GIS or web geocoding tools. Geocoding provides mappable coordinates for your addresses, allowing it to be used in GIS and other mapping software.

Global positioning systems (GPS). A series of satellites relays signals used by a GPS receiver to determine the location of the device. Inexpensive GPS receivers provide location and elevation. More sophisticated devices allow you to append attribute information (data at the location) and export the data so that it can be mapped in GIS.

Cell phones. Cell phones can generate approximate locations using your situation relative to cell towers. Cell phones may supplement these data by using wi-fi and GPS signals. Smart phone applications allow you to use your phone to collect locations and attributes at those locations.

Data collection on existing maps. It's easy to collect primary data by marking them on an existing map. Researchers used a city property map to record the location of electrical poles and powerlines. Such data can be digitized, scanned, or added (by hand) to, and used with, existing layers of data in GIS and other mapping software.

Remote sensing imagery. Images of the earth, taken from airplanes or satellites (remote sensing devices). Imagery is available at different resolutions and can include non-visible energy such as ultraviolet. Imagery can also be used to generate mappable data: roads can be traced or vegetation types delineated. Remotely sensed imagery is often combined with other map data in GIS.

Crowdsourcing. Websites can collect mappable data from anyone who can access the site and enter information. Thousands of users from around the world are constructing the OpenStreetMap.org open source map of the world. In essence, the "crowd" is the data source.

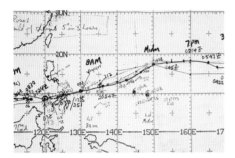

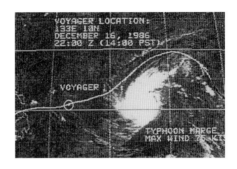

Primary data for *The Flight of Voyager*.
Primary data for the Voyager map were provided by Len Snellman and Larry Burch, the two meteorologists responsible for monitoring weather conditions for the flight. The data consisted of annotated maps, data tables, and satellite imagery.

Snellman and Burch hand-compiled, on maps (left, top three maps), detailed data including a series of flight-path locations (latitude and longitude) for Voyager with associated wind direction, wind velocity, airplane altitude, and time.

DiBiase and Krygier created a table of relevant data (left, bottom) to guide the mapping of the Voyager path and associated flight data.

Weather information was documented in satellite images (above) that served as the basis for the depiction of storms on the final Voyager map.

Gathering data for your map can take a lot of time.

Secondary Data Sources

Secondary data are derived from primary data: aggregations of traffic counts, generalizations of vegetation types.

Scanned and digitized paper maps
Federal government data, including census, economic, environmental data
Regional and local government data, including detailed road, property, and zoning data
For-profit data providers
Public domain data providers, such as GeoCommons
Non-governmental organizations such as the UN and World Bank

Tertiary Data Sources

Secondary data can be assembled in turn; thus resulting in tertiary data. If maps made from primary data are evidence, and maps made from secondary data are like reports of evidence, maps made from tertiary data would be akin to indices of law cases.

Maps are often made from other maps. Map makers don't think about all this assemblage as generalization, but it is. Each step away from the phenomena makes the map less and less about the phenomena and more and more about the data.

Secondary data sources for the Voyager map consisted of a basic Mercator map projection (taken from a printed paper map), place names from an atlas, and sunrise and sunset information from an ephemeris (below).

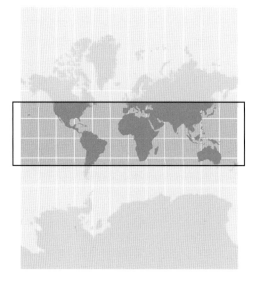

Interpreting Data

Making maps requires other maps, thinking, and interpretation. Erwin Raisz's fabulous landform map of Mexico (excerpt, bottom) began with air photos and topographic maps, like the map with interpretive annotations by Raisz (top). Raisz reviewed and interpreted the data he had – contours and images – and from them created caricatures of the key landforms in the region. Raisz's interpretations resulted in a map that is more than the sum of its original data sources.

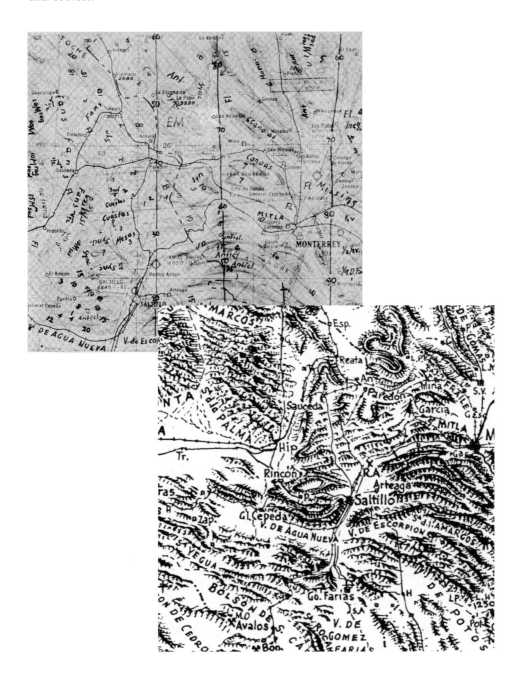

Data Organization

At a basic level, mappable data can be organized into two broad categories, as either qualitative (differences in kind) or quantitative (differences in amount). Such data distinctions guide analysis and map symbolization.

Qualitative

Differences in kind. Also called nominal data.

House and business locations
Rivers and lakes
Electoral college wins, by state (Democrats or Republicans)
Dominant race in a block-by-block map of a town
Location of different bird species seen in a nature preserve

Symbolization: shown with symbols, pictographs, or icons; or with differences in color hue (red, green, blue), as such colors are different in kind, like the data.

Quantitative

Differences in amount: includes ordinal, interval, and ratio data.

Estimated number of same-sex couples, living together, in the U.S. (by county)
Total number of Hispanics in a block-by-block map of a town
Number of loggerhead shrikes counted in a nature preserve

Symbolization: shown with differences in color value (dark red, red, light red), as such colors suggest more and less, like the data.

Levels of quantification

Ordinal: order with no measurable difference between values.
   Low-, medium-, and high-risk zones

Interval: measurable difference between values, but no absolute zero.
   Temperature Fahrenheit (30° is not half as warm as 60°)

Ratio: measurable difference between values, with an absolute zero value.
   Total population in countries

Things are complicated!

And unduly so! The difference between streets and alleys in the Boylan Heights neighborhood (left) is *qualitative* (alleys serve a different purpose than streets) and *quantitative* (streets are larger and carry more traffic). The map symbols reflect this complexity.

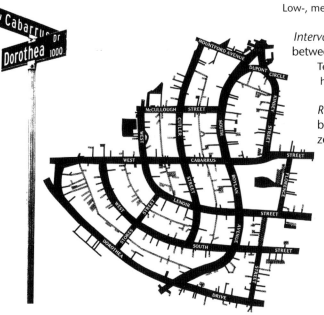

Digital Data Organization

In the 21st century qualitative or quantitative data are likely to be digital. There are two basic ways that digital geographic data are organized and stored: vector or raster. Vector data consist of points that can be connected into lines, or areas. Raster data consist of a grid of cells, each with a particular value or values.

Vector Data

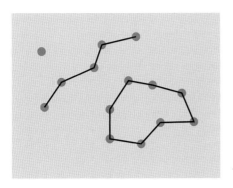

Raster Data

Vector data consist of located points (nodes), lines (a connected series of points), and areas (a closed, connected series of points, also called polygons). Attribute information can be appended to a point, line, or area and stored in a related database. A line standing for a road includes attributes such as name, width, surface, etc. Design characteristics can be appended to points, lines, and areas.

Sources and use: GPS devices collect vector data; many public and private sources of mappable data (USGS, Census TIGER, KML/KMZ) provide data in vector format. Common GIS software uses vector format data. Graphic design software, such as *Illustrator* or *Corel Draw,* also use vector data, making the conversion of GIS output into graphic design software relatively easy.

Raster data consist of a grid with values associated with each grid cell. Higher-resolution raster files have smaller cells. Remotely sensed imagery is raster: each cell contains a level of energy reflected or radiated from the earth in the area covered by the cell. Raster data can have points (one cell), lines (a series of adjacent cells), or areas (a closed series of adjacent cells). Raster data can also include attributes.

Sources and use: Common raster data include satellite and aerial imagery available from public and private sources. Most GIS software allows you to use raster and vector data together. GIS software, such as the open-source *GRASS,* works with raster data. Image editing software, such as *Photoshop* or the open-source *GIMP,* use raster data and can import raster GIS output.

Transforming Data

Raw data, whether primary or secondary, may need to be transformed in order to make a map maker's point. It may be more useful to use totals instead of individual instances; it could make more sense to report phenomena as so many per unit area; an average temperature might be more meaningful than a bunch of daily highs and lows; or if your point has to do with change, rates might be helpful. There is always a *motivation* behind data transformations; choose wisely for an effective map.

Total Numbers

The total number of some phenomenon associated with a point, line, or area.

Amount of pesticide in a well
Pounds of road kill collected in a county

Symbolization: Variation in point size or line width. Represent whole numbers in areas with a scaled symbol for each area.

Densities

The number of some phenomenon per unit area. Divide the number of people in a country by the square area of the country.

Doctors per square km in a country
Adult bookstores per square mile in U.S. cities

Symbolization: Variation in color lightness and darkness in the areas.

Averages

Add all values together and divide by the number of values in the data set. Can be associated with points, lines, and areas.

Average monthly rainfall at a weather station
Average age of murder victims in U.S. cities

Symbolization: Variation in point size or line width. Variation in color lightness and darkness in areas.

Rates

The number of some phenomenon per unit time. May be associated with points, lines, or areas.

Cars per hour along a road
Murders per day in major cities

Symbolization: Variation in point size or line width. Variation in color lightness or darkness in areas.

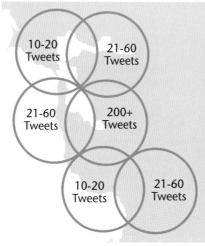

Tweeting "earthquake": The U.S. Geological Survey collects the number of Twitter messages with the word "earthquake" to assess the location of earthquakes around San Francisco, California. Total numbers can be misleading, as many more people live in the area with the highest number of tweets. Instead, transform the data into the percent of tweets per total population.

Time and Data

Map makers talk about space as though the spatial location of a phenomenon were key to its understanding. Location is important, but no more so than *when* it occurred. Every phenomenon happens at some place and at some time. Many phenomena change over time, and single maps or sequences of maps at regular intervals can effectively reveal spatial and temporal patterns. Such a sequence naturally suggests animation. But not all sequences require animation.

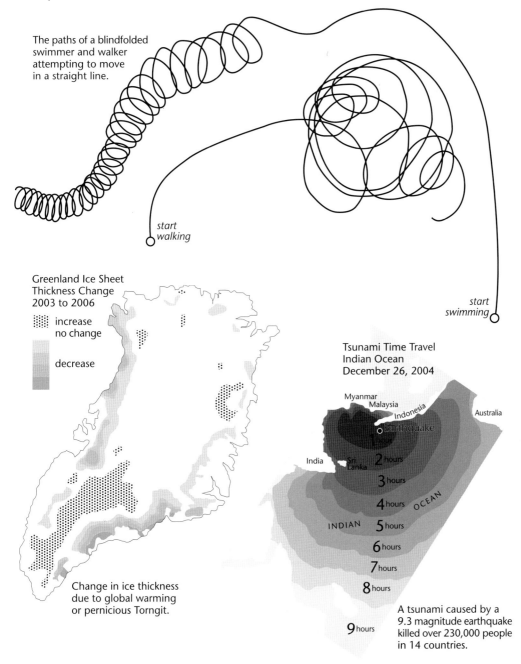

The paths of a blindfolded swimmer and walker attempting to move in a straight line.

start walking

start swimming

Greenland Ice Sheet Thickness Change 2003 to 2006

increase
no change

decrease

Change in ice thickness due to global warming or pernicious Torngit.

Tsunami Time Travel
Indian Ocean
December 26, 2004

Myanmar
Malaysia
Indonesia
Australia
Earthquake
India
Sri Lanka
1
2 hours
3 hours
4 hours   OCEAN
INDIAN   5 hours
6 hours
7 hours
8 hours
9 hours

A tsunami caused by a 9.3 magnitude earthquake killed over 230,000 people in 14 countries.

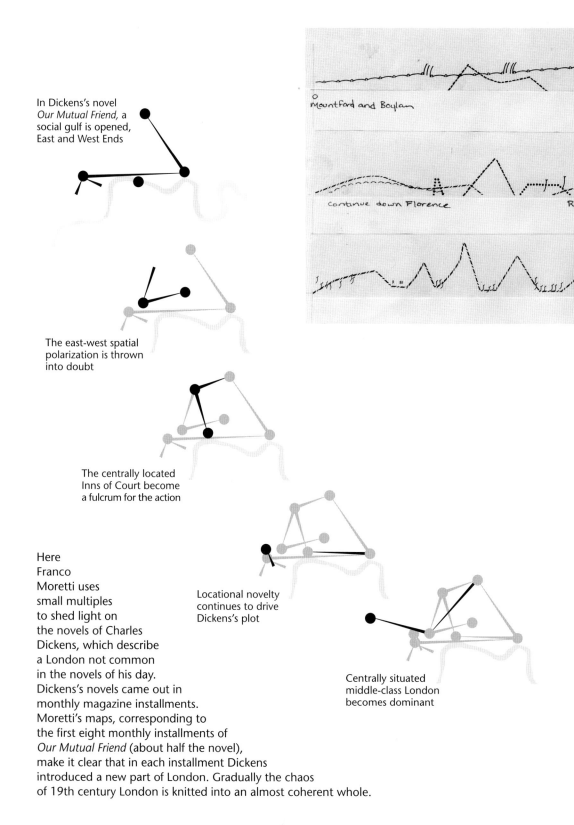

In Dickens's novel *Our Mutual Friend,* a social gulf is opened, East and West Ends

Mountford and Boylan

continue down Florence

The east-west spatial polarization is thrown into doubt

The centrally located Inns of Court become a fulcrum for the action

Here Franco Moretti uses small multiples to shed light on the novels of Charles Dickens, which describe a London not common in the novels of his day. Dickens's novels came out in monthly magazine installments. Moretti's maps, corresponding to the first eight monthly installments of *Our Mutual Friend* (about half the novel), make it clear that in each installment Dickens introduced a new part of London. Gradually the chaos of 19th century London is knitted into an almost coherent whole.

Locational novelty continues to drive Dickens's plot

Centrally situated middle-class London becomes dominant

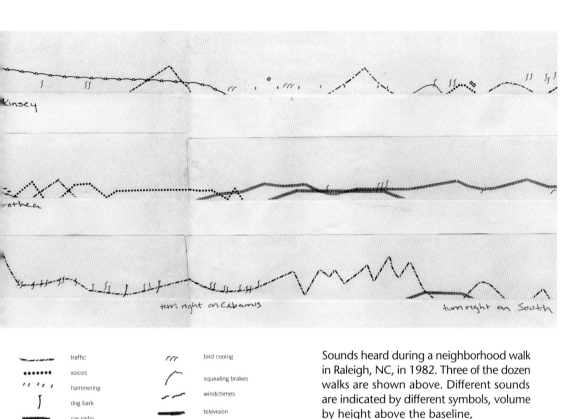

Kinsey

rothea

turn right on Cabarrus   *turn right on South*

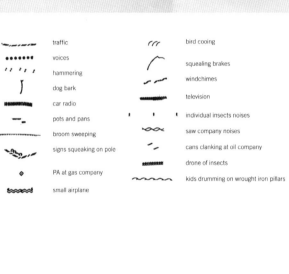

～～～	traffic	⌒⌒⌒	bird cooing
••••••	voices	⌒	squealing brakes
⸴ ⸴ ⸴ ⸴	hammering	～	windchimes
J	dog bark	▬▬▬	television
▬▬▬▬▬	car radio	⸴ ⸴ ⸴	individual insects noises
～─	pots and pans	⌁⌁	saw company noises
⸴⸴⸴⸴⸴⸴⸴⸴⸴⸴	broom sweeping	⸴⸴	cans clanking at oil company
～～	signs squeaking on pole	▬▬▬▬	drone of insects
◇	PA at gas company	～～～	kids drumming on wrought iron pillars
～～～～	small airplane		

Sounds heard during a neighborhood walk in Raleigh, NC, in 1982. Three of the dozen walks are shown above. Different sounds are indicated by different symbols, volume by height above the baseline, time in horizontal distance.

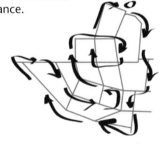

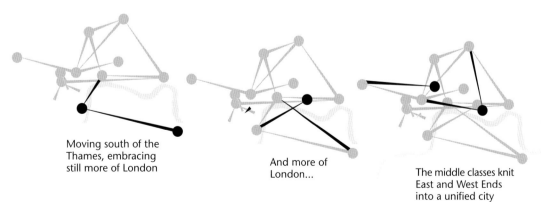

Moving south of the Thames, embracing still more of London

And more of London...

The middle classes knit East and West Ends into a unified city

53

Data Accuracy

There are many types of accuracy associated with data and maps. One approach to accuracy is to ask a series of questions about your data.

Ways to Think about Data Accuracy

Are the facts accurate?

An Israeli tourism ad on the London Underground included a map showing the occupied West Bank, the Gaza Strip, and the Golan Heights as parts of Israel. They're not.

Are things where they should be?

The National Imagery and Mapping Agency put the Chinese Embassy in the wrong location on a map of Belgrade, Yugoslavia. In 1999 a NATO jet accidentally bombed the Embassy, killing 3 and injuring 20.

Does detail vary across the data set?

Wisconsin sand and gravel data combine detailed data from some counties (east side of map) with crude data from other counties (west side of map).

When were the data collected?

U.S. Census data are collected every 10 years; thus a map made in 2009 using 2000 U.S. Census data is using 9 year-old data.

What are the assumptions behind the data?

106
Avg.
IQ
59

People with a high intelligence quotient tend to believe IQ is a valid way to assess human intelligence. Always consider the assumptions that shape the creation of mappable data.

Are the data from a trustworthy source?

Just who says Sara Quick is a bitch? From the website *Bad Neighbor Map,* with its crowd-sourced data.

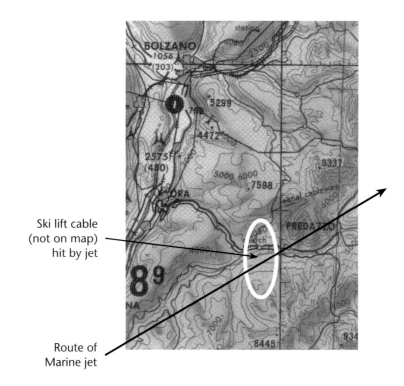

Ski lift cable (not on map) hit by jet

Route of Marine jet

20 Killed Near Cavalese, Northern Italy

Despite a long tradition of map making and expensive new technologies for making maps, maps may still fail and kill.

A U.S. Marine Corps jet struck and severed a ski lift cable spanning an Alpine valley in northern Italy in 1998, sending the ski lift gondola crashing to the earth, killing 20 people.

The jet crew did have a map of the area, but not one that showed the ski lift cable. The cable was shown on Italian maps, but the Pentagon prohibits the use of maps made by foreign nations.

Map users and makers must always be critical of maps. Maps have the potential for failure that every human-created object has.

Digital Data

Widespread use of geographic information systems and the development of extensive databases of digital GIS data require an understanding of metadata and copyright.

Metadata

Metadata are data about data. Dependable digital geographic data include such detailed information as

Identification information: general description of the data

Quality information: which defines the data quality standards

Spatial data organization information: how spatial information in the data is represented

Spatial reference information: coordinate and projection information

Entity and attribute information: map data and associated attributes

Distribution information: data creator, distributor, and use policy

Metadata reference information: metadata creator

Citation information: how to cite information when used

Temporal information: when data were collected, updated

Contact information: how to contact data creator

Metadata standards in the U.S. have been set by the Federal Geographic Data Committee (www.fgdc.gov). Geographic data providers often follow such guidelines.

Copyright

Copyright is a form of protection provided by the laws of a country to the creators of original works. Intended to reasonably compensate them for their efforts, it was originally of limited duration and included exceptions such as the fair use doctrine. The exceptions have been weakened over time, making current copyright laws more burdensome. In the U.S., copyrights include:

Reproduction of copies of the original copyrighted work

Preparation of derivative works based on the original copyrighted work

Distribution/sale/transfer of ownership of the original copyrighted work

Maps, globes, and charts are covered under U.S. copyright law. This copyright does not extend to the data, the facts themselves. What is copyrighted is the representation of the facts – the line weights, the colors, the particular symbols – as long as this representation includes an "appreciable" amount of original material. So, you can make a map based on the data included on a copyrighted map, but you can't photographically reproduce it.

Works created since 1978 are automatically copyrighted, and there's no way to tell if something is copyrighted or not by looking at it (unless it has a notation to that effect). Given this opacity, it's best to assume that works are copyrighted until you learn otherwise.

Copyleft

Copyleft refers to an array of licensing options encouraging reuse, reproduction, distribution of, and modifications to creative works within certain parameters. Copyleft counters the restrictions and prohibitions of copyright in that only "some" rather than "all" rights are reserved by the creator of a work. Copyleft strategies are integral to the philosophy behind open-source software and collectively created, crowdsourced data (such as Wikipedia).

The GNU General Public License is used for open-source software. For example, the raster GIS software GRASS is available under a GNU license.

Creative Commons Licenses are used for other creative works, including maps, geographic data (OpenStreetMap) and reproductions of historical maps (David Rumsey Collection). Six licenses are offered under Creative Commons 3.0, including the *Attribution-ShareAlike License:*

You are free: to share – to copy, distribute, and transmit the work; to remix – to adapt the work, under the following conditions:

Attribution: You must attribute the work in the manner specified by the author or licensor (but not in any way that suggests that they endorse you or your use of the work).

Share Alike: If you alter, transform, or build upon this work, you may distribute the resulting work only under the same, similar, or a compatible license.

Public Domain

Creative works and content neither owned nor controlled by anyone are said to be in the public domain. Public domain works may be used by anyone for any purpose without restriction.

Anything copyrighted in the U.S. prior to 1923 has now entered the public domain, and everything copyrighted is subject to the fair use doctrine. As a general rule of thumb, federal government maps and data in the U.S. are in the public domain.

Public domain and copyleft licensed maps and geographic data are a great idea when making maps. Consider using copyleft licensing on maps and geographic data you create. Creativecommons.org provides easy methods for licensing your work.

I wanna hang a map of the world in my house. Then I'm gonna put pins into all the locations that I've traveled to. But first I'm gonna have to travel to the top two corners of the map so it won't fall down.

Mitch Hedberg (no date)

The *Atlas* maps, writing, and illustrations were done by people who live in thatch-roof, wooden houses they made themselves and who eat food they grew themselves. They got up early in the dark morning hours to make wood fires to cook tortillas and warm coffee before walking to their milpas to cultivate corn and beans, and then mapped their fields, rain-forest hunting grounds, traditional medicine places, and ancient ruins.

Maya Atlas (1997)

Ignorance exists in the map, not in the territory.

Eliezer Yudkowsky, *Mysterious Answers to Mysterious Questions* (2007)

More...

Data is a topic of much interest among philosophers of science. What are data? How are they created? How are they used? The literature's vast, and an interesting place to start is Bruno Latour's *Pandora's Hope: Essays on the Reality of Science Studies* (Harvard University Press, 1999). For a review of GIS, mapping, and data, try David DiBiase's open-source web textbook *The Nature of Geographic Information* (www.e-education.psu.edu/natureofgeoinfo). Also see Paul Longley, Michael Goodchild, David Maguire, and David Rhind, *Geographic Information Systems and Science* (Wiley, 2010); Michael Zeiler, *Modeling Our World* (ESRI Press, 1999); and David Arctur and Michael Zeiler *Designing Geodatabases* (ESRI Press, 2004). For *very* different perspectives on mappable data, see some of the books listed at the end of the preceding chapter, Leslie's *A Sense of Place,* Tobias's *Chief Kerry's Moose,* Wood's *Everything Sings,* and Thompson's *Experimental Geography.*

There are lots of sites on the web for free mappable data, such as *Natural Earth* (naturalearthdata.com). For literary mapping and time see Franco Moretti, *Atlas of the European Novel* (Verso,1998) and *Graphs, Maps, Trees* (Verso, 2005). The *Creative Commons* web pages have super resources on copyright and copyleft licensing. Search for "map copyright" on the web for a diversity of materials on the subject.

Sources: Simon Elvins, *FM Radio Map* courtesy of the artist (simonelvins.com). Tree fall map redrawn from map created by Backwoods Orienteering Klub (backwoodsOK.org). Pin oak map from E.N. Munns, *Distribution of Important Forest Trees of the United States* (U.S. Department of Agriculture, Washington, 1938). Vegetation zones map redrawn from *Goode's World Atlas* (Chicago, 1990, p. 16). Map symbols for nautical dangers from "Section O of Chart #1," *Nautical Chart Symbols and Abbreviations* (Washington DC, 1957). Erwin Raisz's Mexico maps from his "A New Landform Map of Mexico," *International Yearbook of Cartography* 1, 1961 (plates between p. 124-125). Boylan Heights street and alley map from Denis Wood, *Everything Sings: Maps for a Narrative Atlas*, Siglio Press, 2010. Delaware County, Ohio, air imagery courtesy of the DALIS Project, Delaware County, Ohio (dalisproject.org). Earthquake tweet data from the U.S. Geological Survey's *Twitter Earthquake Detector (TED).* Blind walker and swimmer data from Emily Davis, 1928, "Why Lost People Go in Circles" *The Science News-Letter* 14:378, pp. 3-4. Greenland ice sheet change data from NASA's *Earth Observatory* website (earthobservatory.nasa.gov). Indian Ocean tsunami data from the National Oceanic and Atmospheric Administration (noaa.gov). Charles Dickens's *Our Mutual Friend* data from Franco Moretti, *Atlas of the European Novel* (Verso,1998). Boylan Heights neighborhood sound diagram from Denis Wood, *Everything Sings: Maps for a Narrative Atlas* (Siglio Press, 2010). NATO Belgrade map from NATO (nato.int). Wisconsin sand and gravel map based on *Glacial Deposits of Wisconsin* map (Wisconsin Geological and Natural History Survey, 1976). U.S. Census map of Delaware, Ohio, from the U.S. Census Bureau's *American Factfinder* (factfinder.census.gov). IQ map data from *Wikipedia Commons* (wikipedia.org). The Cavalese map is courtesy of the Harvard Map Library, Harvard University.

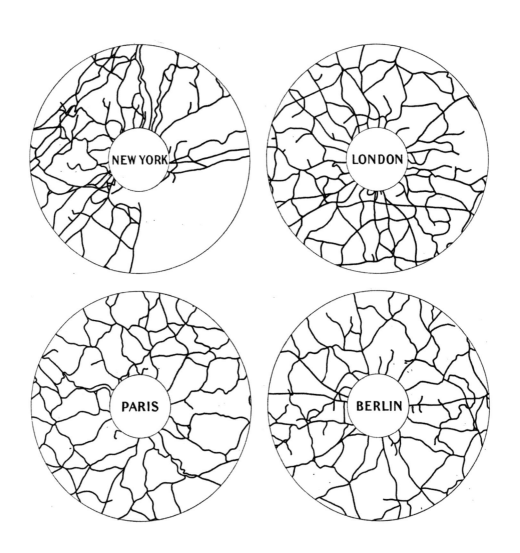

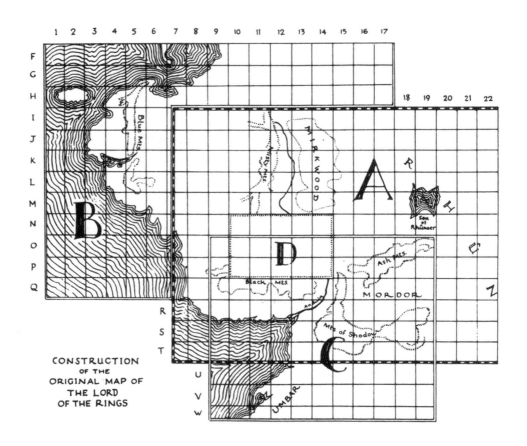

CONSTRUCTION
OF THE
ORIGINAL MAP OF
THE LORD
OF THE RINGS

How is it made?

The extension of the map (B) added the coastlands necessary for the conclusion of *The Return of the King.*

J.R.R. Tolkien's son, Christopher, made this diagram of the structure of his father's continuously evolving map of Middle Earth. Sheet A was the first to be drawn, and includes the lands mapped for the elder Tolkien's first book, *The Hobbit.*

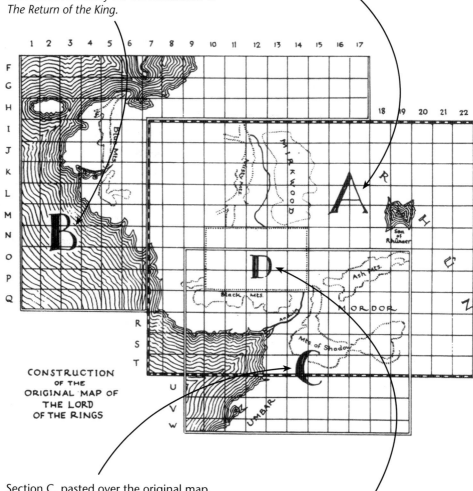

CONSTRUCTION
OF THE
ORIGINAL MAP OF
THE LORD
OF THE RINGS

Section C, pasted over the original map with a strong glue, replaced Tolkien's original, sketchy versions for the setting of *The Two Towers* and *The Return of the King.* The elder Tolkien made his maps with colored inks, pencil, and colored chalk.

The final part (D) of the evolving map was replaced again and again as Tolkien wrote and rewrote the first half of *The Two Towers.* Tolkien folded and unfolded the collaged map so often that the crease marks are hard to distinguish from the joints between the pieces.

4 Map Making Tools

The reason a map is being made suggests the appropriate tools. Even if J.R.R. Tolkien had known how to use GIS software, it would not have been any better than the pencils, paper, and glue he used for the maps he made while working on his *Lord of the Rings* trilogy. The map of Middle Earth was iteratively developed as Tolkien wrote his books, and consists of various pieces superimposed and glued together over time. Pencils, ink, chalk, paper, and glue are great map-making and thinking tools. So too are internet mapping sites and GIS software. Choose appropriate tools based on what you need to do.

Making Maps without Computers

You certainly don't need a computer to make maps. Indeed, map making with pencils and paper is appropriate, inexpensive, and effective in many instances. A sketch map made with pencils and paper may be your final map, or it may be a vital step in the process of producing a map with other tools. J.R.R. Tolkien's hand-drawn map of Middle Earth helped him envision plots and settings in his trilogy. Visual thinking and discovery are not limited to any specific map-making tools.

Old-school mapping tools. Back when the Voyager map was made (1987), the map-making tools of choice were lighted drafting tables, scribers, technical pens, peel-coat film, and stick-up type (right). Within five years such tools would be replaced by computers, software, digitizing tables, scanners, and mice. Map-making tools change quickly, but map design principles should transcend these changes.

Empower without power. Computer mapping doesn't work when you don't have a computer. Sketch mapping, here in the Philippines, engages community members in compiling cultural and economic resources. In many parts of the world, computers, computer skills, and electricity are simply not an option. Sketch maps are as useful as their computer counterparts, and certainly may be digitized for use on the computer.

Win the election. Volunteers for the Democratic presidential candidate go door to door in a key city in a swing state. Using a map printed from an internet site and a pen, they mark their opponent's supporters as an X and their supporters as an O. A filled O means a supporter who may not vote (such as an elderly person with no transportation). Addresses and phone numbers of these supporters are entered into a spreadsheet. These people will be called on election day to urge them to vote and to offer free transport to the polls.

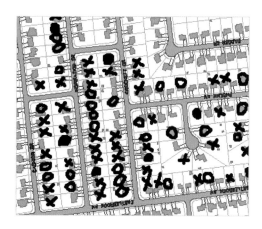

66

Sketch and discover. Abraham Verghese used maps to help think about his HIV-infected patients. Dr. Verghese practiced medicine in rural Tennessee. He and his colleagues were stunned when HIV-infected patients began to dominate their practices. What was this urban problem doing in rural Tennessee? "There was a pattern in my HIV practice. I kept feeling if I could concentrate hard enough, step back, and look carefully, I could draw a kind of blue-print that explained what was happening here..." Dr. Verghese borrowed a map of the U.S. from his son. With the map spread on his living room floor, he marked where his HIV patients lived. He labeled the map Domicile, but he could as well have called it "Birthplace," for most of his patients were men who had come home to die.

Dr. Verghese next mapped where his HIV patients lived between 1979 and 1985. The places on the Acquisition map "seemed to circle the periphery of the United States" and were mostly large cities. "As I neared the end, I could see a distinct pattern of dots emerging on this larger map of the USA. All evening I had been on the threshold of seeing. Now I understood." Dr. Verghese learned of a circuitous voyage, a migration from home and a return, ending in death. It was "the story of how a generation of young men, raised to self-hatred, had risen above the definitions that their society and upbringing had used to define them."

The maps Verghese made on his living room floor with pencils and paper might not be much to look at, but the thinking they inspired was invaluable.

Domicile

Acquisition

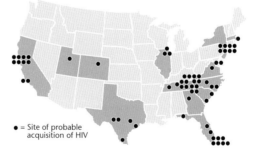

Making Maps on the Internet

The internet has democratized map making – at least for the 22% or so of people in the world with internet access. Anyone with access to the internet can grab mappable data and make maps on sophisticated interactive mapping sites. Such sites are usually free and easy to use but limited in functions, compared to mapping software you can purchase. Internet mapping sites are increasingly sophisticated, with more mapping functions and more control over map design.

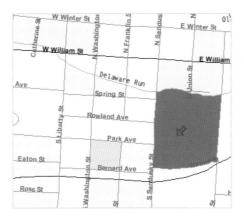

Statistical data with location information (place names, postal codes, addresses, etc.) allows the data to be mapped by internet and desktop mapping tools. Check out the GeoCommons, The Geography Network, or any national census bureau. Always check for metadata and use data critically. The 1990 U.S. Census assigns the 1490 students living on campus at Ohio Wesleyan University to a block where only one person lives.

Order paper maps on the net. You can order lots of stuff on the internet, including ingestible clay, special sausages from Spain, and paper maps showing the world or your neighborhood. Such maps are portable and cheap and an excellent source for making new maps, for plotting and thinking about your data, or for inspiration. Check out national topographic mapping agencies, private map dealers, or online auction sites.

Static maps on the net. Static digital copies of historical and contemporary maps are available from map libraries, government sources, and commercial sites. Such maps are useful as sources for making maps, reference, and illustration. Static maps are typically in raster format without any geographic coordinates. Check out David Rumsey's vast collection of historical maps, or just do a internet image search with a place name and "map."

Locations, directions, peeking at places. Many internet sites allow you to map locations and routes or to peek at an image of your house. Such sites replace the need to sketch directions by hand or use a road map. The maps at these sites may also be useful as a data source for making maps. MapQuest, Yahoo! Maps, and Google Maps are but a few of the many free commercial sites where you can make maps of locations and routes.

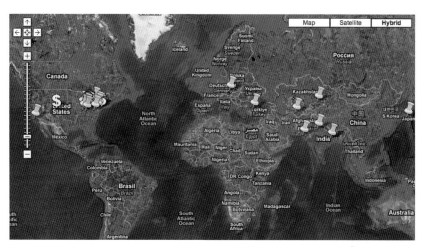

Map mashups combine mappable data sources stored in different places. Mashups allow you to map your own data over maps from Google, Yahoo, etc., using an API (application programming interface). This Google Maps mashup shows the homes of students in an introductory mapping course at Ohio Wesleyan.

Cloud-computing map applications. Map design and creation on the internet using software and data stored at other internet sites, often charging fees based on the resources used. Platform-independent cloud-computing mapping tools, such as *Indiemapper,* provide control over map projections, data classification, type, color, layout and include diverse import/export capabilities.

GIS analysis. Delaware County (Ohio) DALIS site allows you to search for a particular property in the county, then create a "buffer" around that property. For example, you may ask the site to show all properties within 100 feet of Richard Fusch's property. Such functions are very useful if Dr. Fusch needs to inform neighbors within a set distance from his property of impending construction.

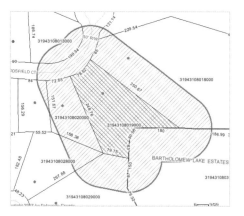

Making Maps with GIS

Geographic information systems are a collection of technologies for gathering, processing, analyzing, and mapping geographic information. The analysis of geographic data includes queries (show all parcels owned by the city), buffers (show all roads within 100 feet of wetlands), and overlay (combine a map of soils with a map of wetlands, and show only the soils that underlie wetlands). In many cases, GIS users interact with maps while engaged in GIS analysis, and the outcome of the analysis is a map. Because of this, most GIS software has map making and map design capabilities.

A collaboration among students, planners, and community members uses GIS to plan for trail development in Delaware County, Ohio. GIS data (roads, property parcels, railroads, etc.) are acquired. *Queries* to the parcels layer produces a map of all schools (flagged symbols) and parks (gray areas). Potential trails should connect schools and parks, and this map helps generate potential trail corridors that do just that.

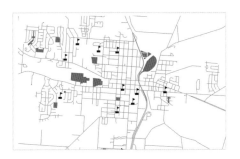

Adjacent property owners sometimes object to rail-to-trail conversions. A 600-foot *buffer* is made around the rail grades. *Overlay* the buffers on the parcels layer to determine the percent of adjacent residential, commercial, and industrial properties. Rail grades with more adjacent residential property may arouse more concern. This map helps to anticipate potential conflict over rail-to-trail development.

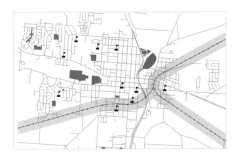

Collaborators use GIS analytical tools to generate potential trails that maximize connectivity between schools and parks but minimize potential conflict. To this "working map" in GIS is added potential trail routes, collected with GPS. This map embodies the process of using GIS and maps to generate a speculative network of recreational trails.

It is time to go public with the map. Schools and parks are labeled, and roads added. The railroad is interested in selling the rail grade to the city, so that potential trail is left on the map. Another potential trail is adjacent to property owners who may object, so it is removed from the map. The map is designed to look professional, suggesting that the process of generating potential trails was professional.

Making Maps with Graphic Design Tools

Internet mapping tools offer limited control over map design, and the results are typically crude raster files. GIS mapping tools offer more control over map design and publication-quality output but still may not meet the needs of those who produce high-quality maps for publication. Graphic design software offers extensive control over text, color, and all point, line, and area symbols on a map. While not designed with map making in mind, graphic design software can import internet or GIS files, re-create and redesign them, and generate files that can be professionally printed.

The U.S. Environmental Protection Agency's Enviromapper internet site makes maps locating toxic emissions, storage, and contamination. Super information, but the map doesn't work for a project mapping toxins around Indianola Informal Alternative School in Columbus, Ohio. The toxic sites and schools are not easy to see, and there are too many roads. Save the map from the internet as a raster (.jpg) file and import it into graphic design software.

The redesigned map better suits the needs of the community group interested in drawing attention to toxic release sites near schools. The toxic site and school symbols are larger, distracting details are left off the map, and the zip code area is easier to see. Graphic design software provides extensive control over map design and helps create more effective maps.

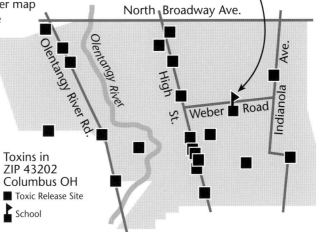

Indianola Informal Alternative School
251 Weber Road

Toxins in ZIP 43202 Columbus OH
■ Toxic Release Site
▶ School

Making maps requires many tools

Envision the best map possible based on what you need your map to do, and harness different software to make your vision work:
No single software package can do everything!

Whereas maps, like guns, must be accurate, they have the additional advantages that they are inexpensive, don't require a permit, can be openly carried and used...

Bernard Nietschmann, "Defending the Reefs" (1995)

...The Indians are very expert in delineating countries upon bark, with wood coal mixed with bear's grease, and which even the women do with great precision.

John Long, *Voyages and Travels* (1791)

A little instruction in the elements of chartography – a little practice in the use of the compass and the spirit level, a topographical map of the town common, an excursion with a road map – would have given me a fat round earth in place of my paper ghost.

Mary Antin, *The Promised Land* (1912)

[Cyrus has made a map in the dirt by using various bits of junk to represent buildings.]
Cyrus: Considering my audience, I'm gonna make this very quick and very simple.
   [points with a stick] This is the boneyard. This is the hangar. This is our plane.
Viking: What's that? [points at a rock]
Cyrus: That's a rock.
Viking: O-okay.

Conair (1997)

Kim: You suck at drawing, don't you?
Scott: Maps are hard! I could draw it really good if it was a sheep.

Bryan Lee O'Malley, *Scott Pilgrim vs. the World* (2005)

More...

Older map making texts are all about making maps with feet, eyeballs, pens, and paper. A classic, very much still being used, is David Greenhood's *Down to Earth: Mapping for Everybody* (Holiday House, 1944) republished as *Mapping* (University of Chicago Press, 1964). Few of the maps in Wood's *Everything Sings* were made with a computer (the atlas provides production notes for all its maps). Few parish maps are made with computers.

While not necessary, most of you will be making your maps on computers. Billions and billions of manuals and guides and websites explain how to use GIS software, and web mapping sites are usually easy to use without much guidance. Thoughtful GIS books include Nicholas Chrisman, *Exploring Geographical Information Systems* (Wiley, 2001) and Francis Harvey, *A Primer of GIS* (Guilford Press, 2008), and Paul Longley, Michael Goodchild, David Maguire, and David Rhind, *Geographic Information Systems and Science* (Wiley, 2010).

A neat introduction to graphic design software tools is the *Digital Foundations* website (wiki.digital-foundations.net) and book by Xtine Burrough and Michael Mandiberg (AIGA Design Press, 2009).

For an engaging overview of maps and graphs in action, helping to figure stuff out, see Howard Wainer, *Graphic Discovery* (Princeton University Press, 2005).

Mark Monmonier's thought-provoking books are full of stories about how maps do their work in the world (all University of Chicago Press): *Cartographies of Danger: Mapping Hazards in America* (1997), *Air Apparent: How Meteorologists Learned to Map, Predict, and Dramatize Weather* (1999), *Bushmanders and Bullwinkles: How Politicians Manipulate Electronic Maps and Census Data to Win Elections* (2001), *Spying with Maps: Surveillance Technologies and the Future of Privacy* (2002), *From Squaw Tit to Whorehouse Meadow: How Maps Name, Claim, and Inflame* (2007), *Coast Lines: How Mapmakers Frame the World and Chart Environmental Change* (2008), and *No Dig, No Fly, No Go: How Maps Restrict and Control* (2010).

Sources: The Tolkien map is from J.R.R and C. Tolkien, *The History of Middle Earth,* vol. 3 (Houghton Mifflin and HarperCollins Publishers, 1989). Sketch mapping photo courtesy of *Participatory Avenues* website (iapad.org). Dr. Verghese's AIDS data from Abraham Verghese, *My Own Country: A Doctor's Story* (Vintage, 1994) and "Urbs in Rure: Human Immunodeficiency Virus Infection in Rural Tennessee" (*Journal of Infectious Diseases,* 160:6). *Indiemapper* (indiemapper.com) image courtesy of Axis Maps (axismaps.com). Google *My Maps* mashups courtesy of Google. Delaware County, Ohio property maps courtesy of the *DALIS Project,* Delaware County, Ohio (dalisproject.org). Toxic map for zip code 43202 redrawn from the U.S. Environmental Protection Agency's *Enviromapper* website.

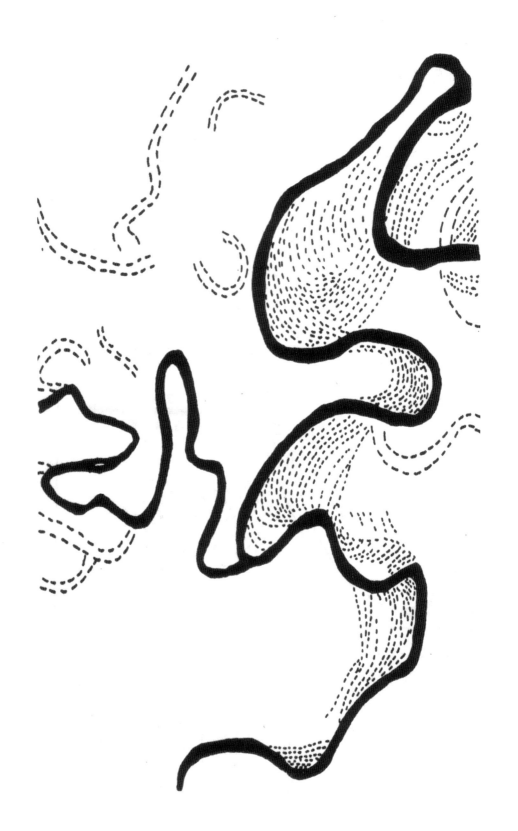

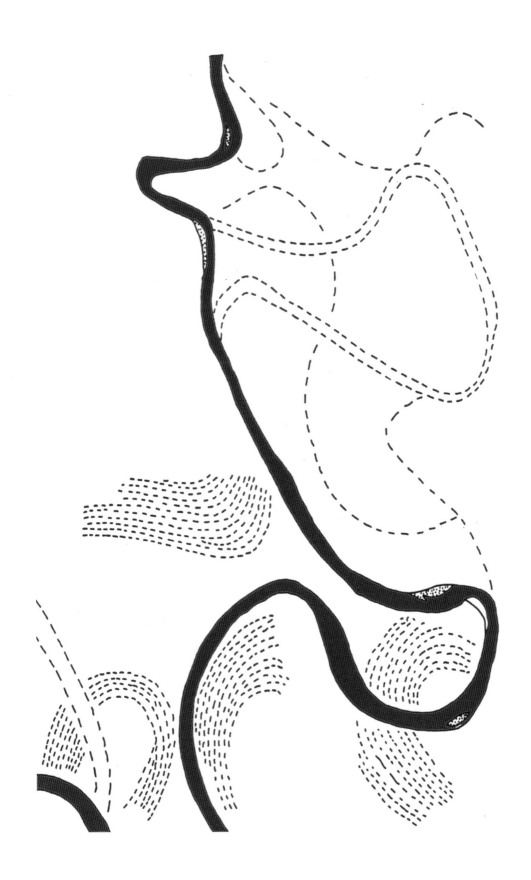

How do you flatten,
shrink, and locate data?

5 Geographic Framework

We have to mess with our spherical earth to get it flat. Through a process called "map projection" the curved surface of the earth is flattened. We are used to seeing the flat earth on maps, but what if we flattened a human body? Artists Lilla Locurto and Bill Outcault scanned the entire surface of their bodies with a 3D object scanner. They brought the data into *GeoCart* map projection software and projected it. We have clearly become accustomed to the radical transformation involved in projecting the earth.

Map Projections

Our earth's surface is curved. Most maps are flat. Transforming the curved surface to a flat surface is called map projection. Projected maps are flat, compact, portable, useful, and always distorted. Any curved surface gets distorted when you flatten it.

An orange peel *tears* when you peel and flatten it.

A toad skin *tears* when you peel and flatten it.

Strike flat the thick rotundity o' th' world!

William Shakespeare, *King Lear*

The surface of the earth *tears* when you peel and flatten it. Peel a globe and you'll get globe gores (below).

Most map projections stretch and distort the earth to "fill in" the tears. The Mercator projection (bottom) preserves angles, and so shapes in limited areas, but it greatly distorts sizes. Note the size of Greenland on the globe as compared to the Mercator.

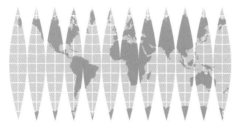

Thinking about Map Projections

There are many ways to think about map projections, beyond flattened bodies, oranges, and toads. How is the earth's surface distorted? How do map projections distort your data? What do particular map projections preserve from the spherical earth?

Distorting Circles

In the 19th century, Nicolas Auguste Tissot developed his "indicatrix," which is used to evaluate map projection distortion. Imagine perfect circles of the same size placed at regular intervals on the curved surface of the earth. These circles are then projected along with the earth's surface. Distortions in the area and shape (angles) of the circles show the location and quality of distortions on the projected map.

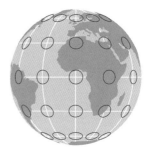

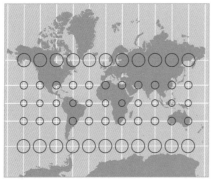

Mercator map projection:
Preserves shapes, distorts areas.

Tissot's circles change area as you move north and south of the equator on the Mercator map projection (left). The more distorted the circles, the more distorted the areas of the land masses. Circle shapes are not distorted.

Tissot's circles change shape over the surface of this area-preserving map (below). The more distorted the circles, the more distorted the shapes of the land masses. Circle areas are not distorted.

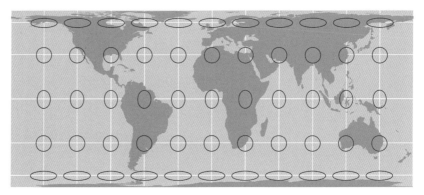

Equal-area map projection: Preserves areas, distorts shapes.

Distorting Data

Mappable data are always associated with a location on the earth's surface. That is, mappable data are always tied to the grid. Because this grid gets distorted when it's projected from the curved surface of the earth to the flat surface of the map, the data tied to the grid are distorted too. Projections matter because of what they do to our data! It's important that what map projections do to our data clarifies, not muddies, it.

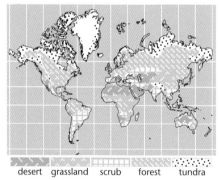

| desert | grassland | scrub | forest | tundra |

Mercator map projection:
Preserves shapes, distorts areas.

A map of vegetation on a Mercator (left) distorts the data. Northern types are greatly expanded in area compared to those near the equator. This suggests the global dominance of northern vegetation types, which is wrong.

The same data on a map projection (below) that doesn't distort areas. But now shape is distorted! You must be smart with projections and understand tradeoffs. In this case, with area vegetation data, you are better to distort shapes and not areas.

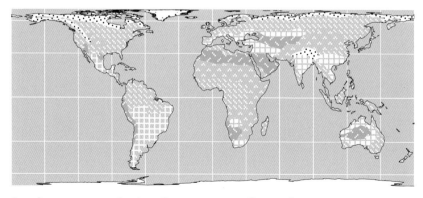

Equal-area map projection: Preserves areas, distorts shapes.

What Map Projections Preserve

No map projection preserves the attributes of a globe, which maintains area, shape, distance, and direction. Map projections can preserve one or two of the attributes of the globe, but not all at once. Select a map projection that makes the best sense for your data.

Preserving Area

Some projections preserve area. This means that areas which are the same on the globe are the same on the map. Area-preserving (equal-area) map projections, are a good default in particular for maps showing area data.

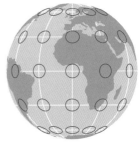

Mollweide projection: Oval shape, preserves area. Rounded map shape suggests the round earth. The Mollweide can be recentered to minimize shape distortions of regions of greatest interest.

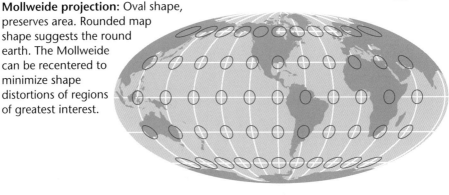

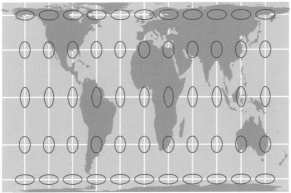

Peters (Gall-Peters) projection: To some map experts, what garlic is to vampires. This area-preserving projection's straight grid makes north-south relationships straight forward. As with any rectangular map projection, it fits into page layouts.

This is a good projection for illustrating the shape distortions inherent in area-preserving map projections and the area distortions inherent in shape-preserving projections. Also good as a symbol of affinity with the Global South.

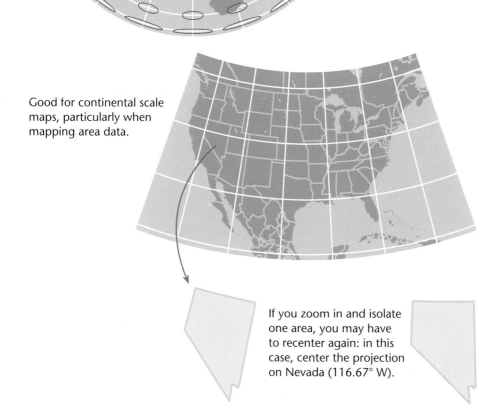

Albers equal-area projection:
A common area-preserving map projection. Poor for world scale maps because of shape distortion and peculiar form. However, recentering on an area of interest (the U.S., below) and selecting part of the earth (continent, country) results in an equal-area map with minimal shape distortion.

Good for continental scale maps, particularly when mapping area data.

If you zoom in and isolate one area, you may have to recenter again: in this case, center the projection on Nevada (116.67° W).

Preserving Shape (Angles)

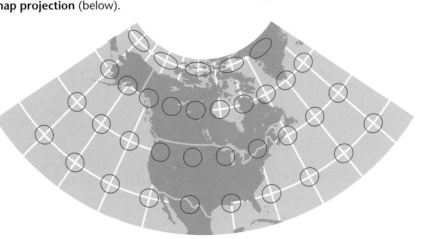

Saying that certain map projections preserve shape is not technically correct, but makes sense to normal people. As long as you are away from areas of high distortion, the shapes of continents look OK in comparison to shapes on the globe. Conformal map projections preserve angles (around points) and therefore shape in small areas. Consequently, sizes are distorted. Conformal map projections are good for mapping regions and continents, especially when statistical data are involved. At sub-global scales distortions of area and shape are not as evident.

Compare the distortion circles on the **Lambert conformal map projection** (right) and **Lambert equal-area map projection** (below).

Mercator projection: One of the few conformal world projections. Its distortions of sizes are nasty, and it is a poor choice for a world map. Good for equatorial maps (below) where the area distortion is small. A Mercator with the distorted north and south lopped off was chosen for the Voyager map. Also good for maps of very small areas (below, right). Many detailed topographic maps are based on the (transverse) Mercator.

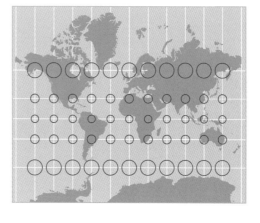

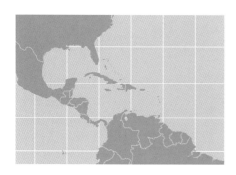

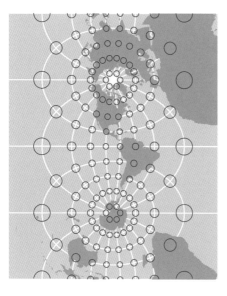

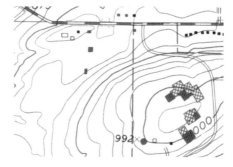

Transverse Mercator: On the Mercator projection, scale is true along the equator. When that projection is recentered sideways along a meridian (or line of longitude), scale is true along that meridian. This recentered projection is known as the Transverse Mercator and is the basis for the Universal Transverse Mercator coordinate system. When areas a few square miles in size are mapped using this projection, they are effectively free of all distortion.

Preserving Distance, Direction

Stretch a piece of string between two points on a globe, and you will get the shortest distance between the points (a great circle). Some projections preserve such distance relations: a straight line between two points on the map is the shortest distance between those two points on the earth. Distance relations cannot be preserved on equal-area maps. Direction can be preserved on area-, shape-, or distance-preserving map projections.

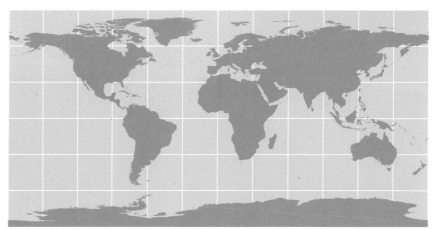

The Geographic Coordinate System, aka the **Geographic** or **Equirectangular projection**, is similar to the **Plate Carrée projection**. Invented by Marinus of Tyre in 100 AD, it "maps meridians to equally spaced vertical straight lines, and circles of latitude to evenly spread horizontal straight lines." It is a common default projection for GIS software and internet mapping sites, where it distorts the data mapped on it.

This unpleasant item preserves nothing but distance, principally for north-south measurements. It is used as a default for computational purposes. Geographic location and image pixels are related in a simple manner that makes drawing and redrawing world maps very quick. Because it is so pervasive, we are growing used to its insidious lies and distortions.

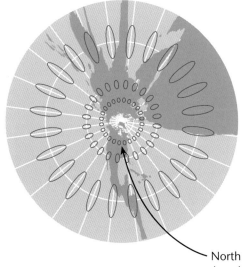

Gnomonic projection: A straight line anywhere on a Gnomonic projection is a great circle route, the shortest distance between two points. Terrifying distortions of area and shape and the inability to show more than half the earth at a time limit other uses of this projection.

North America

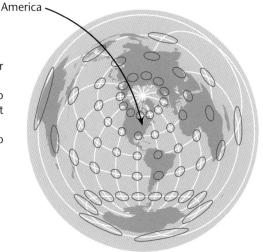

Azimuthal equidistant projection: Planar (azimuthal) map projections preserve directions (azimuths) from their center to all other points. The azimuthal equidistant projection also preserves distance. A straight line from the projection center to any other point represents accurate distance in addition to correct direction and the shortest route. Great for travel agencies when centered on their city.

Poor for showing anything but distances from a point. Areas and shapes are wildly exaggerated as you move from the center.

Preserving Interruptions

Globe gores, peeled from a globe and flattened, are akin to interrupted map projections. Interrupted map projections minimize distortions on the uninterrupted part of the map, and are typically used on maps of the entire earth.

Interrupted map projections are commonly used for maps of global statistical data. They are also used as icons (Berghaus "star") and as the basis of "cut and assemble" do-it-yourself globes (Fuller).

The Berghaus is equidistant north of the equator. The Fuller has constant scale along the edges of all 20 of the triangular pieces. Within each of these triangles, area and shape are well preserved.

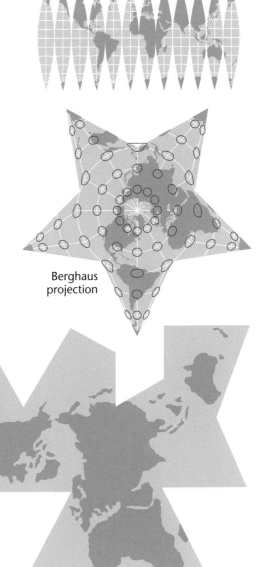

Berghaus
projection

Fuller
projection

Goode's homolosine projection: Goode's is a common interrupted map projection used for world maps of statistical data. The projection does not distort areas, and shape distortions in the uninterrupted areas of the map are minimized.

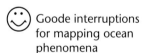 Poor interruptions for mapping ocean phenomena

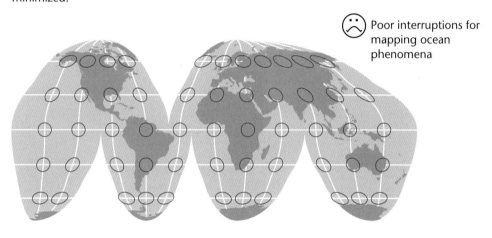

Interruptions can be moved. A map for ocean phenomena can interrupt land areas, for example.

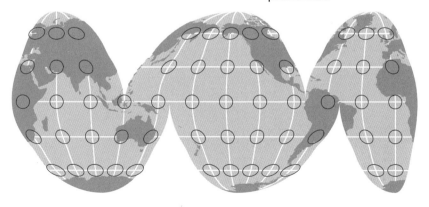 Goode interruptions for mapping ocean phenomena

Preserving Everything, Almost

Map projection is most visible at a global scale, where distortions of areas and shape are most evident. Area-preserving projections often badly distort shapes; and shape-preserving projections, area. But there is an alternative – a map projection that does not distort anything!

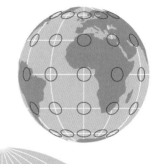

But that doesn't exist. Crap.

However, instead consider some compromise map projections, that distort both area and shape a bit, but neither too badly. They preserve everything, almost.

The Van der Grinten projection does not preserve shape or area, but minimizes their distortions in all but polar regions. Usually the polar regions are lopped off and the map presented as a rectangle.

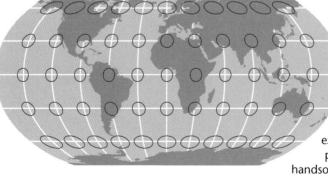

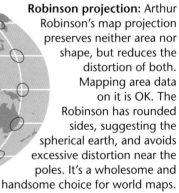

Robinson projection: Arthur Robinson's map projection preserves neither area nor shape, but reduces the distortion of both. Mapping area data on it is OK. The Robinson has rounded sides, suggesting the spherical earth, and avoids excessive distortion near the poles. It's a wholesome and handsome choice for world maps.

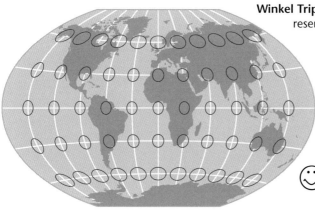

Winkel Tripel projection: This projection resembles the Robinson, but it has less area exaggeration in the polar regions. The Winkel Tripel is the default map projection of the National Geographic Society.

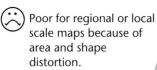 Good for a general world map and for mapping global phenomena.

Poor for regional or local scale maps because of area and shape distortion.

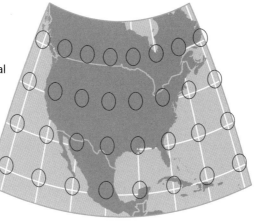

Map Scale

The earth is big. Maps are small. Map scale describes the difference, verbally, visually, or with numbers. Map scale will be determined by your goals for your map. Map scale affects how much of the earth, and how much detail, can be shown on a map.

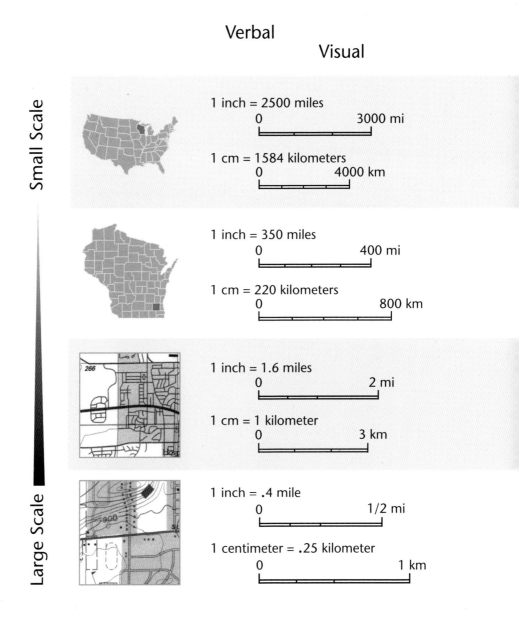

Verbal

Visual

Small Scale

1 inch = 2500 miles
0   3000 mi

1 cm = 1584 kilometers
0   4000 km

1 inch = 350 miles
0   400 mi

1 cm = 220 kilometers
0   800 km

1 inch = 1.6 miles
0   2 mi

1 cm = 1 kilometer
0   3 km

1 inch = .4 mile
0   1/2 mi

1 centimeter = .25 kilometer
0   1 km

Large Scale

Numerical

1 : 155,000,000

A representative fraction (RF) shows the proportion between map distance and earth distance for any unit of measure. 1 inch on the map is 155 million inches on the earth. 1 cm on the map is 155 million cm on the earth.

1 : 22,000,000

Distortions from map projections become less visually noticeable at regional and local scales. These distortions may become evident when combining map layers with different projections in GIS.

1 : 100,000

Divide a representative fraction:

| 1 / 100,000 | = .00001 |
| 1 / 24,000 | = .00004 |

The former is smaller than the latter: thus 1:100,000 is smaller scale than 1:24,000 (larger scale).

1 : 24,000

Larger-scale maps show more detail, but of a limited area. Map projection distortions are less evident and distance is relatively accurate over the entire map.

Small Scale

Large Scale

Earth's Shape and Georeferencing

The earth is a geoid, an imperfect 3D object, more akin to an ellipsoid, a squished-down sphere. Locating stuff we want to map requires a system for locating (georeferencing), typically based on a pair of coordinates. Latitude and longitude are based on our 3D earth. Other systems operate in a projected 2D world.

Ellipsoids and Datums

There are a diversity of ellipsoids that can serve as an approximation of our earth. Many were created to best fit the earth in specific countries. Maps and data sets based on different ellipsoids will not work together.

A datum is based on a single common point shared by the geoid (our imperfect earth) and a particular ellipsoid. All spatial relationships (locations, directions, scales) derive from this single point. Maps and data sets based on different datums will not work together.

Ellipsoids, datums, projections, and georeferencing systems combine in strange ways with great diversity. Expect to convert your geographic data sets in order to get them to work together.

The World Geodetic System of 1984 (WGS84) is the modern standard ellipsoid. The North American Datum of 1983 (NAD83), closely approximating WGS84, has been widely adopted in North America.

Map Coordinates

Map coordinates – also known as georeferences – typically consist of a pair of numbers or letters that locate data, tying them to the grid. Geographic data are distinguished from other data by the fact that they can be located. There are many different map coordinate systems and means of georeferencing. You need to pay attention to which system is associated with your data. Coordinate systems can be converted from one to another, and often have to be when making maps with GIS.

Where is (0, 0)? Where, on earth, should the origin (0, 0) be? If Washington, DC is the origin, then all other locations are in relation to Washington. Different coordinate systems have different origins.

Area covered? How much of the earth is covered by the coordinate system? Coordinate systems may cover all or only part of the earth.

Flat or spherical? Coordinate systems covering part of the earth assume a flat earth, to take advantage of easier planar geometry. Coordinate systems covering the entire earth assume spherical geometry.

Units? Coordinate systems can be in English units (feet), metric units (meters), or degrees. Different coordinate systems have different units of measurement.

Latitude and Longitude

Latitude and longitude cover the entire earth with one system and a single origin. It's used when you need a single coordinate system for our 3D earth. In this system, locations are specified in degrees, which can be subdivided. There are 60 minutes in 1 degree and 60 seconds in 1 minute. Decimal degrees are increasingly common.

The equator is the origin for **latitude.** Lines of latitude are called parallels. Parallels run east-west, measuring 90° north and 90° south of the equator. Parallels never converge: one degree of latitude is always 69 miles or 111 km.

Greenwich, England, is the origin (prime meridian) for **longitude.** Lines of longitude are called meridians. Meridians run north-south, measuring 180° east and 180° west of the prime meridian. Meridians converge at the poles. One degree of longitude at the equator is 69 miles or 111 kilometers. One degree of longitude at the poles is 0 miles or km (a point).

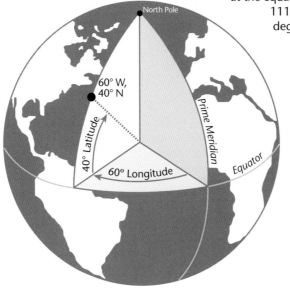

The single origin (0, 0) is off the coast of Africa. Coordinates fall into one of four quadrants to the N/S (latitude) and E/W (longitude) of this origin.

Latitude and longitude can operate in a 3D or 2D world. Map projections flatten and distort the grid of latitude and longitude.

Universal Transverse Mercator (UTM)

The Universal Transverse Mercator
(UTM) is a projected coordinate system.
UTM, based on the "transverse" (sideways)
Mercator projection, covers most of the
earth, which is divided into 60 zones, each
6° wide, running from 84° north to 80°
south. Planar geometry (a flat earth) makes
computations easy. UTM is measured in
meters. A point is located in terms of how
many meters east and north it is from the
origin. UTM is used by environmental
scientists, the military, and any other
professionals who work at a regional or
local scale but need their maps to
coordinate with maps of other areas on
the earth.

Zone 17

UTM zones are widest at the equator and
narrow toward the poles. The poles use
the universal stereographic coordinate
system. Ten UTM zones (10 to 19) cover
the continental U.S.

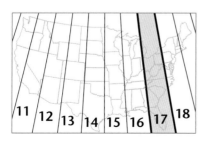

Each 6°-wide zone has a north and south
zone. The central meridian of each zone
is assigned a value of 500,000 meters (not
tied to any earth feature; thus east-west
measurements are called *false eastings*).
North-south measurements are in terms
of the equator. In the south zone, the
equator is assigned a value of 10,000,000
meters to avoid negative coordinates.

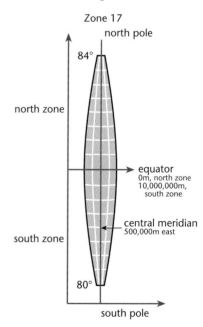

State Plane Coordinate System (SPCS)

The **State Plane Coordinate System** (SPCS) is also based on a flat 2D earth. SPCS is used only in the United States, which is divided into over a hundred areas, each with its own coordinate system. Since each area is relatively small, distortions from projection are minimal. The most recent version of SPCS is based on the NAD83 datum, but not all states have converted.

SPCS is measured in feet, meters, or both. A point is located in terms of how many units east and north it is from the origin. Where the false origin is set varies. Some states use a central meridian as a *false easting,* as with UTM. Others establish the false origin outside of the bounds of the state (but only coordinates within state boundaries are used).

SPCS is used by planners, urban utilities, and environmental engineers. Similar coordinate systems are used in other parts of the world.

Small U.S. states have a single SPCS zone; larger states (excepting Montana) are divided into several zones. No SPCS zone crosses a state boundary.

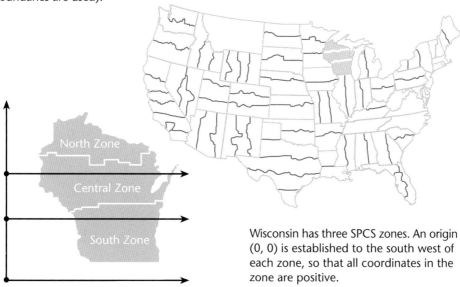

Wisconsin has three SPCS zones. An origin (0, 0) is established to the south west of each zone, so that all coordinates in the zone are positive.

Interrupting my train of thought
Lines of longitude and latitude
Define and refine my altitude

Wire, *Map Ref. 41°N 93°W* (1984)

He could jazz up the map-reading class by having a full-size color photograph of Betty Grable in a bathing suit, with a co-ordinate grid system laid over it. The instructor could point to different parts of her and say, "Give me the co-ordinates."... The Major could see every unit in the Army using his idea.... Hot dog!

Norman Mailer, *The Naked and the Dead* (1948)

According to the map we've only gone 4 inches.

Harry Dunne, *Dumb and Dumber* (1994)

Men can read maps better than women 'cause only the male mind could conceive of one inch equaling a hundred miles.

Roseanne Barr

More...

A highly accessible introduction to map projections is Denis Wood, Ward Kaiser, and Bob Abramms, *Seeing through Maps: The Power of Images to Shape Our World View* (ODT, 2006).

The best book on map projections is John Snyder's *Flattening the Earth* (University of Chicago Press, 1997). Snyder's approach is historical and exhaustive. The treatment is technical but not intimidating.

A great handbook describing and illustrating dozens of map projections is *An Album of Map Projections,* published by John Snyder and Philip Voxland (U.S. Geological Survey Professional Paper #1453, 1989). See also Tau Rho Alpha, Janis Detterman, and James Morley, *Atlas of Oblique Maps* (U.S. Geological Survey Miscellaneous Investigations Series #I-1799, 1988) for a collection of very cool oblique maps. For map projections in the context of GIS, see Jonathan Iliffe and Roger Lott, *Datums and Map Projections: For Remote Sensing, GIS and Surveying* (Whittles Publishing, 2008).

Every cartography textbook has a chapter or two on map projections and coordinate systems. Check out any of the previously cited cartography texts for more information than you will ever need.

Sources: Body projection image courtesy of Lilla Locurto and Bill Outcault. The majority of map projections in this chapter were generated in *GeoCart* software, and a few in ESRI's ArcGIS. The flat toad is redrawn from Edward Tufte, *The Visual Display of Quantitative Information* (Graphics Press, 1983). The map of vegetation on the Mercator map projection is redrawn from a map in Anne Spirn's *The Granite Garden: Urban Nature and Human Design* (Basic Books, 1984). The latitude and longitude earth was redrawn from David Greenhood's *Down to Earth: Mapping for Everybody* (Holiday House, 1951). The state plane coordinate and Universal Transverse Mercator diagrams were redrawn from Philip Muehrcke and Juliana Muehrcke, *Map Use* (JP Publications, 1998) and Kraak and Ormeling's *Cartography* (Pearson, 1996).

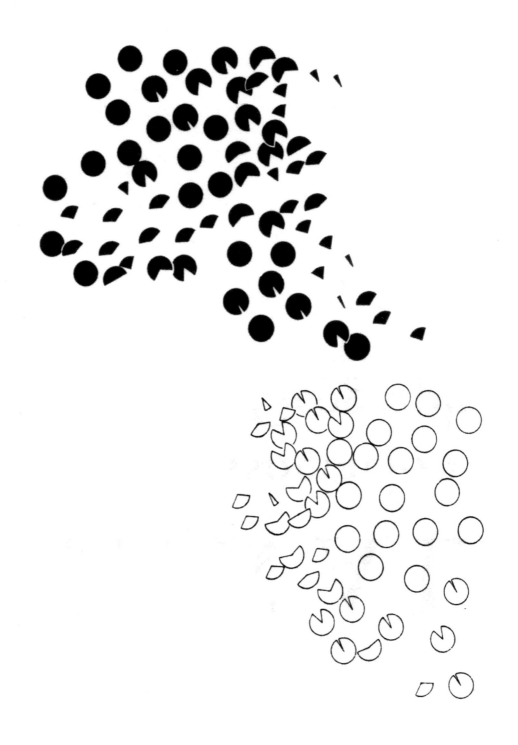

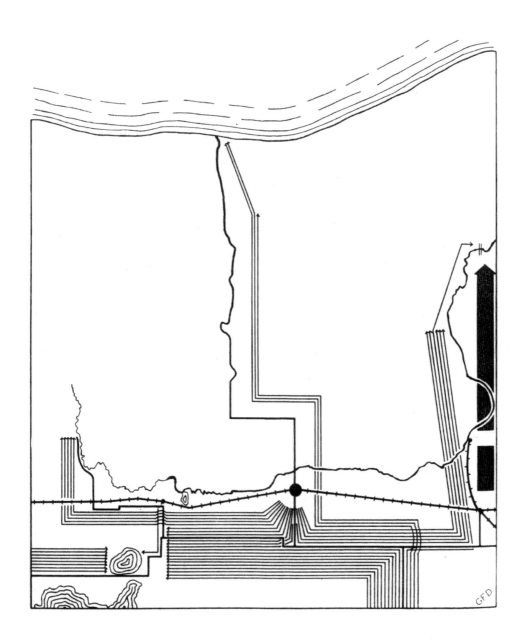

GFD

Uh ... why is the Voyager story *ending* at the *beginning* of the map?

	DAY 9			DAY 8			DAY 7			DAY 6			DAY 5	
Hours Aloft	216 hours	200	192 hours	184	176	168 hours	160	152	144 hours	136	128	120 hours	112	104	96 hours

Fuel on landing: 18 gallons

Triumphant landing at Edwards AFB

WNW

Engine stalled; unable to restart for five harrowing minutes

Transition from tailwinds to headwinds

NNW 20

ENE 18

ESE 14

Oil warning light goes on

Rutan disabled by exhaustion

Thunderstorm forces Voyager into 90° bank

E 37

E 34

Passing between two mountains, Rutan and Yeager weep with relief at having survived Africa's storms

E 20

Gabon

Flying among 'the redwoods': life and death struggle to avoid towering thunderstorms

Discovery of backwards fuel flow

Worried about flying through restricted airspace, Rutan and Yeager mistake the morning star for a hostile aircraft

E 10-20

Cooler seal lea

Squall line

United States

Atlantic Ocean

Pacific Ocean

Nicaragua

Costa Rica

NW 10-15

Cameroon

Congo Zaire

Tanzania

Kenya

Uganda

Ethiopia

Somalia

Atlantic Ocean

Visibility

Altitude (feet): 20,000 / 15,000 / 10,000 / 5,000 / sea level

Distance: 26,678 miles traveled | 5,000 miles to go | 10,000 miles to go / 12,532 miles previous record

Flight data courtesy of Len Snellman and Larry Burch, Voyager meteorologists
Mapped by David DiBiase and John Krygier, Department of Geography, University of Wisconsin-Madison, 1987

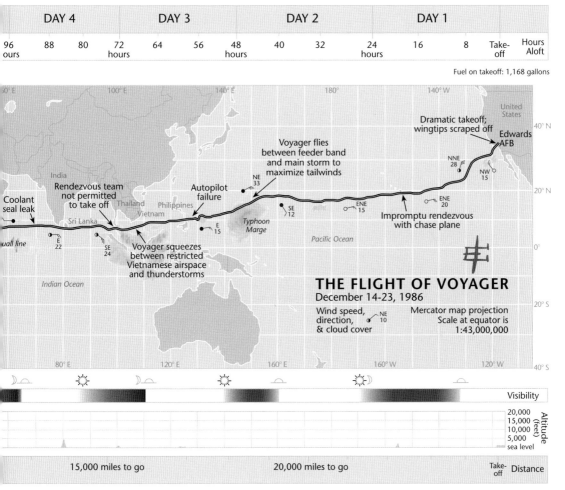

DAY 4				DAY 3			DAY 2			DAY 1			
96 hours	88	80	72 hours	64	56	48 hours	40	32	24 hours	16	8	Take-off	Hours Aloft

Fuel on takeoff: 1,168 gallons

Dramatic takeoff; wingtips scraped off

Edwards AFB

United States

Voyager flies between feeder band and main storm to maximize tailwinds

Autopilot failure

NE 33

NNE 28

NW 15

ENE 20

ENE 15

Impromptu rendezvous with chase plane

SE 12

Rendezvous team not permitted to take off

Coolant seal leak

India

Thailand
Philippines
Vietnam

Sri Lanka

E 15

Typhoon Marge

Pacific Ocean

uall line

E 22

SE 24

Voyager squeezes between restricted Vietnamese airspace and thunderstorms

Indian Ocean

THE FLIGHT OF VOYAGER
December 14-23, 1986

Wind speed, direction, & cloud cover   NE 10

Mercator map projection
Scale at equator is
1:43,000,000

Visibility

Altitude (feet): 20,000 / 15,000 / 10,000 / 5,000 / sea level

15,000 miles to go   20,000 miles to go   Take-off   Distance

Voyager pilots: Dick Rutan and Jeana Yeager
Voyager designer: Burt Rutan

What's up with that?

There was a problem. Voyager's route began in California and headed west, sending the reader backwards across the endpapers and so violating reading habits.

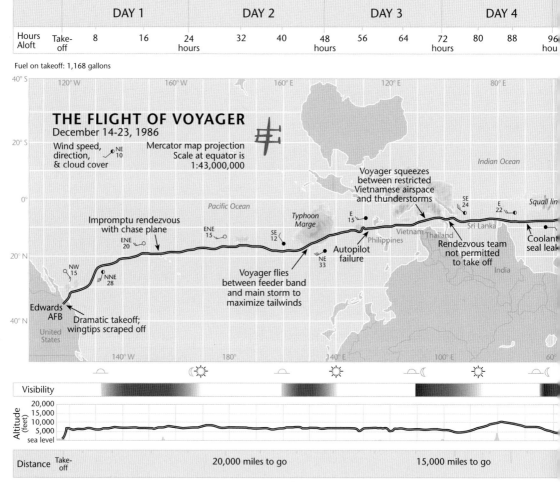

	DAY 1			DAY 2				DAY 3			DAY 4	

| Hours Aloft | Take-off | 8 | 16 | 24 hours | 32 | 40 | 48 hours | 56 | 64 | 72 hours | 80 | 88 | 96 hou |

Fuel on takeoff: 1,168 gallons

THE FLIGHT OF VOYAGER
December 14-23, 1986

Wind speed, direction, & cloud cover

Mercator map projection
Scale at equator is
1:43,000,000

Impromptu rendezvous with chase plane

Voyager flies between feeder band and main storm to maximize tailwinds

Edwards AFB
Dramatic takeoff; wingtips scraped off

Typhoon Marge

Autopilot failure

Voyager squeezes between restricted Vietnamese airspace and thunderstorms

Rendezvous team not permitted to take off

Coolant seal leak

Squall lin

Pacific Ocean

Indian Ocean

Vietnam Thailand
Philippines

Sri Lanka

India

United States

Visibility

Altitude (feet): 20,000 / 15,000 / 10,000 / 5,000 / sea level

Distance Take-off   20,000 miles to go   15,000 miles to go

Flight data courtesy of Len Snellman and Larry Burch, Voyager meteorologists
Mapped by David DiBiase and John Krygier, Department of Geography, University of Wisconsin-Madison 1987

Thankfully, north isn't really up. The data suggest south up, and we suggested south up for the map. "No!" shrieked the publisher. "You can't have south up." And we didn't, in the book. But now, here it is.

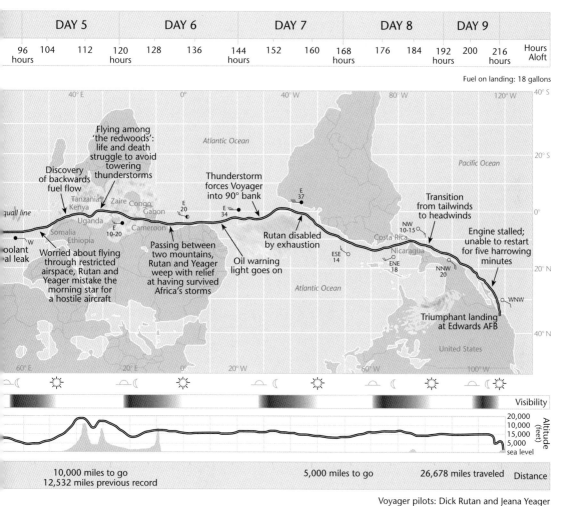

Voyager pilots: Dick Rutan and Jeana Yeager
Voyager designer: Burt Rutan

CHAPTER 6 The Big Picture of Map Design

Map design is tough. What's the point of your map? What kind of data do you have? What tools are you using? What's your geographic framework? With answers to these questions in mind, intelligently design the diverse pieces of your map into a coherent whole. One "big picture" approach to map design is borrowed from advertising: layout and visual arrangement. *The medium is the message.* Another approach is Edward Tufte's "graphical excellence." *The data are the message.* Both approaches played a role in designing the Voyager map, as they should with any mapping project.

Map Pieces

The "big picture" of map design in part requires arranging and rearranging your map with its associated map pieces: title, scale, explanatory text, legend, directional indicator, border, sources and credits, and insets and locator maps. Your intent for the map will determine which are used and how they are displayed on your map.

Title

Titles should, if possible, include

> What: the topic of the map
> Where: the geographic area
> When: temporal information

Title type size, in general, should be two to three times the size of the type on the map itself and bolder. A subtitle, in smaller type, is good for complex map subjects.

 title content:

Population Change

 title size:

Population Change

 title content and size:

Population Change in Ohio
By county, 1900-2000

Scale

Maps from local to continental scale should include a scale, especially if your map's readers need to make measurements on the map. Verbal and visual scales are more intuitive, numerical scales more flexible (they can be metric or English). If your map's users might reduce or increase the size of the map, a visual scale is best (it will remain accurate even if scaled).

⊢——⊢——⊢————⊣ mi
1 inch = 1000 mi
1 : 1,200,000

Small-scale maps (of the entire earth or a substantial portion of it) should not include a simple visual scale, because such maps always contain substantial scale variations.

Explanatory Text

Explanatory text is often vital to the success of your map. You can't express everything you need your map's readers to understand with the map itself. Use text blocks on the map to communicate information about the map content, its broader context, and your goals.

On a historical map, include a paragraph setting the historical context, and also include blocks of text on the map that tell what happened at important locations.

Explain your interpretation of your map's patterns with text: tell your map's readers (in addition to showing them with the map) what you think about the mapped data.

On a choropleth map showing changes in average income over the past decade (by county in a state), explain your interpretation that suburban counties are becoming richer, and urban and rural counties poorer, due to recent tax cuts.

The readers of your map may agree or disagree with your interpretation, but your interpretation and intent will be clearly communicated.

Legend

Map legends vary greatly but should include any map symbol you think may not be familiar to your audience. The legend is the key to interpreting the map.

However, don't insult your map's readers by including obvious symbols in the legend. You need not preface the legend with "Legend" or "Key," as most map readers know that without being told.

Directional Indicator

Use a directional indicator if

The map is not oriented north
The map is of an area unfamiliar to your intended
audience

Directional indicators can often be left off
the map if the orientation is obvious (unless
your audience is stupid). If included, avoid
appallingly large and ugly directional
indicators such as those found in many
GIS software packages.

Border

A border, or neatline, drawn around your
map and its pieces may draw everything
together. Try your map without borders.
It gives a more open feel to the design. If
a border is used, make it narrow, possibly
in gray: noticeable but not distracting.

Sources, Credits, Etc.

Maps may include

Data sources and citations
Map maker and date
Organization and logos
Disclaimers and legal information
Map series information
Copyright and use issues
Map projection and coordinate system

Including projection and coordinate system
information is important if you think
someone may combine your map with
other data in GIS.

Insets and Locator Maps

Choosing a single map scale is difficult. At
a smaller scale, data in one area on the
map are too dense. At a larger scale, the
dense data can be distinguished, but the
geographic extent is limited.

An inset that jumps to a larger scale (less
area, more detail) helps the map viewer
to understand areas on the map where
data are more dense and difficult to
distinguish at the scale of the main map.
An inset that jumps to a smaller scale (more
area, less detail) helps the map viewer
contextualize the area of the main map.
An orthographic map projection is
commonly used as a smaller-scale global
locator map.

Negley Sidebottom Ball:
Significant Wisconsin Buildings

Larger-scale maps benefit greatly by
including a smaller-scale locator map.

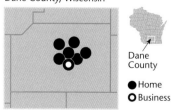

Negley Sidebottom Ball Buildings
Dane County, Wisconsin

109

The **title** succinctly describes the subject and dates of the event. The plane's silhouette visually reiterates the subject. The global context will be clear to the map's audience, and is left out of the title.

The **legend** explains the one symbol on the map that may be unfamiliar to a general audience: the *wind barb*.

Map scale is shown as a representational fraction. The scale on global maps varies significantly, so including a scale can be deceptive. But not including one can be confusing. The compromise is to include a scale with a statement of its limitations.

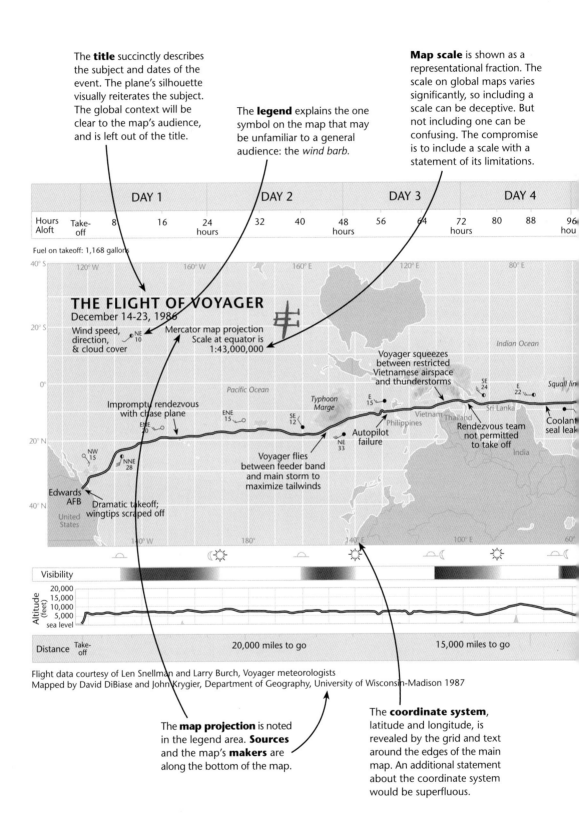

	DAY 1	DAY 2	DAY 3	DAY 4

| Hours Aloft | Take-off | 8 | 16 | 24 hours | 32 | 40 | 48 hours | 56 | 64 | 72 hours | 80 | 88 | 96 hou |

Fuel on takeoff: 1,168 gallons

40° S   120° W   160° W   160° E   120° E   80° E

THE FLIGHT OF VOYAGER
December 14-23, 1986

Wind speed, direction, & cloud cover   Mercator map projection
Scale at equator is 1:43,000,000

Indian Ocean

Voyager squeezes between restricted Vietnamese airspace and thunderstorms

Squall line

Impromptu rendezvous with chase plane

Pacific Ocean

Typhoon Marge

Sri Lanka

Coolant seal leak

Autopilot failure

Rendezvous team not permitted to take off

Philippines   Vietnam Thailand

Voyager flies between feeder band and main storm to maximize tailwinds

India

Edwards AFB

Dramatic takeoff; wingtips scraped off

United States

140° W   180°   140° E   100° E   60°

Visibility

Altitude (feet) 20,000 15,000 10,000 5,000 sea level

Distance   Take-off   20,000 miles to go   15,000 miles to go

Flight data courtesy of Len Snellman and Larry Burch, Voyager meteorologists
Mapped by David DiBiase and John Krygier, Department of Geography, University of Wisconsin-Madison 1987

The **map projection** is noted in the legend area. **Sources** and the map's **makers** are along the bottom of the map.

The **coordinate system**, latitude and longitude, is revealed by the grid and text around the edges of the main map. An additional statement about the coordinate system would be superfluous.

110

Key events are described along the flight path. It's impossible to express this part of the story without **explanatory text**. The text is brief, clear, and tied to the story told in the book.

Inset maps could be used to reveal more detail. In the case of this map, such inset maps were not necessary given the general overview of the flight the map is intended to show.

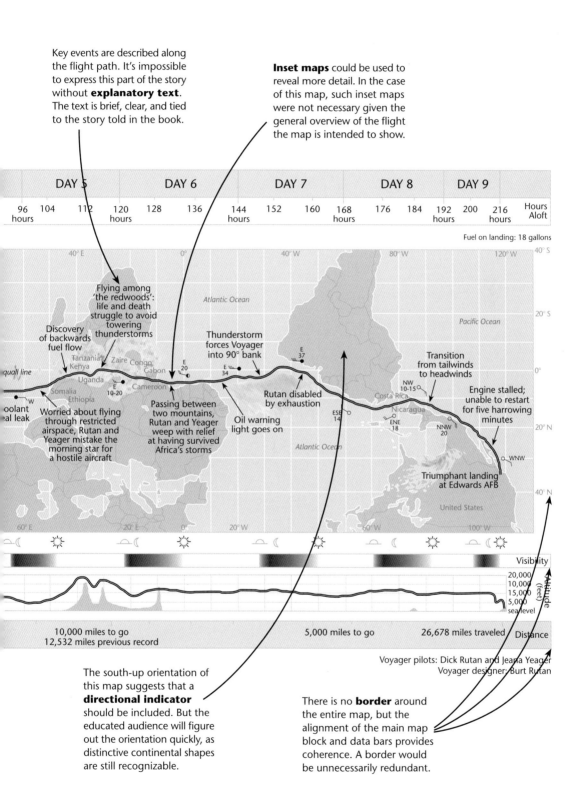

DAY 5	DAY 6	DAY 7	DAY 8	DAY 9	

| 96 hours | 104 | 112 | 120 hours | 128 | 136 | 144 hours | 152 | 160 | 168 hours | 176 | 184 | 192 hours | 200 | 216 hours | Hours Aloft |

Fuel on landing: 18 gallons

40° E   0°   40° W   80° W   120° W   40° S

Atlantic Ocean

20° S

Pacific Ocean

Flying among 'the redwoods': life and death struggle to avoid towering thunderstorms

Discovery of backwards fuel flow

Thunderstorm forces Voyager into 90° bank

Transition from tailwinds to headwinds

Tanzania
Kenya   Zaire Congo
            Gabon   E
squall line   20   E
        Uganda   34
Somalia E   Cameroon
Ethiopia 10-20   E
                                       37
                        Rutan disabled
                        by exhaustion
                                    Costa Rica   NW
                                                 10-15
                              ESE   Nicaragua
                              14   ENE
W   18   NNW
oolant   20
al leak Worried about flying   Oil warning   WNW
         through restricted   light goes on
         airspace, Rutan and
         Yeager mistake the   Passing between
         morning star for   two mountains,
         a hostile aircraft   Rutan and Yeager
                               weep with relief
                               at having survived
                               Africa's storms

Engine stalled; unable to restart for five harrowing minutes

0°

Atlantic Ocean

20° N

Triumphant landing at Edwards AFB

40° N

United States

60° E   20° E   0°   20° W   60° W   100° W

Visibility

20,000
10,000
15,000   Altitude (feet)
5,000
sea level

10,000 miles to go
12,532 miles previous record

5,000 miles to go

26,678 miles traveled   Distance

Voyager pilots: Dick Rutan and Jeana Yeager
Voyager designer: Burt Rutan

The south-up orientation of this map suggests that a **directional indicator** should be included. But the educated audience will figure out the orientation quickly, as distinctive continental shapes are still recognizable.

There is no **border** around the entire map, but the alignment of the main map block and data bars provides coherence. A border would be unnecessarily redundant.

111

Thinking about the Big Picture

Abstract map diagrams like the small ones on the following page assume that map content doesn't matter. Map content matters more than anything. Interesting content trumps bad design every time. But why not strive for both? Diagrams like these can draw your attention to the many ways in which map pieces can be arranged and rearranged to enhance your goals for your map. Tufte's "graphical excellence" serves as a complementary approach, driven by your data, to the big picture of map design.

Roving Eyes

Eyes move over maps, so can we guide the way they move by design? George Jenks pioneered eye movement studies of maps. In 1973 he published the eye scans below. Arrows indicate the start of the path of the subject's eyes over the map. Jenks's study and subsequent research show that people's eye movements over maps are erratic, individualistic, and nearly impossible to predict.

But where there's nothing else to hang on to, assuming a predictable eye movement path is one strategy that can help the map maker focus his or her attention on the way the map and its pieces are coming together. The big picture of map design is experimentation guided by intuition and evaluation. Design strategies include playing with paths, the visual center, balance, symmetry, sight-lines, and grids.

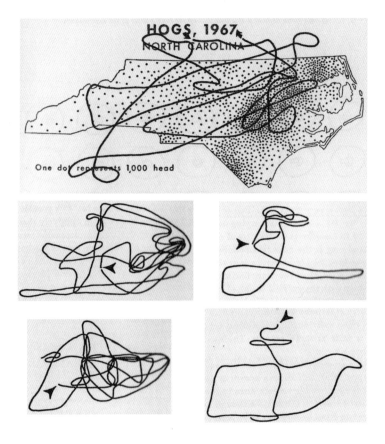

Visual Arrangement

Try assuming that reading a map follows a **path**, like reading a page in a book, from the upper left to the lower right. Arrange map pieces so that those that should be seen first are in the upper left of the map – unless, of course, your language reads from right to left.

The **visual center** of a map is slightly above the actual center. Centering implies importance. Try positioning map pieces so that the most important are near the visual center of the map. The map reader may focus near this center and assume the elements there are most important.

Balance can be assessed when all your map pieces are in place. Map pieces vary in weight: some seem heavier, others lighter. Does your arrangement seem off-balance? Unless you want to suggest a lack of balance to your readers, rearrange your map pieces for better balance.

Symmetry can be thought of as balance around a central vertical axis. Symmetrical balance (left) feels traditional, conservative, and cautious. Asymmetry (right) depends on off-center weights and balances. It typically feels modern, progressive, complex, and more creative.

Sight-lines are invisible horizontal or vertical lines that touch the top, bottom, or sides of map elements. Minimizing the number of sight-lines reduces disjointedness and stabilizes and enhances map layout. This allows your map's readers to focus on the map's subject.

Symmetrical grids (left) are based on two central axes and top, bottom, and side margins. **Asymmetrical grids** (right) are more complex, but still depend on the visual center while maintaining top, bottom, and side margins.

The reader's eyes are drawn to the **upper left** corner of the map, where we typically begin reading text on a page and where this story begins.

With south up, there's a perfect place for the title. The typical reader will read the title, grasp the topic, then set off on the **path** of Voyager, heading to the right and west.

The title and legend are **asymmetrical** on the page, lending a subtle sense of complexity and creativity to the map.

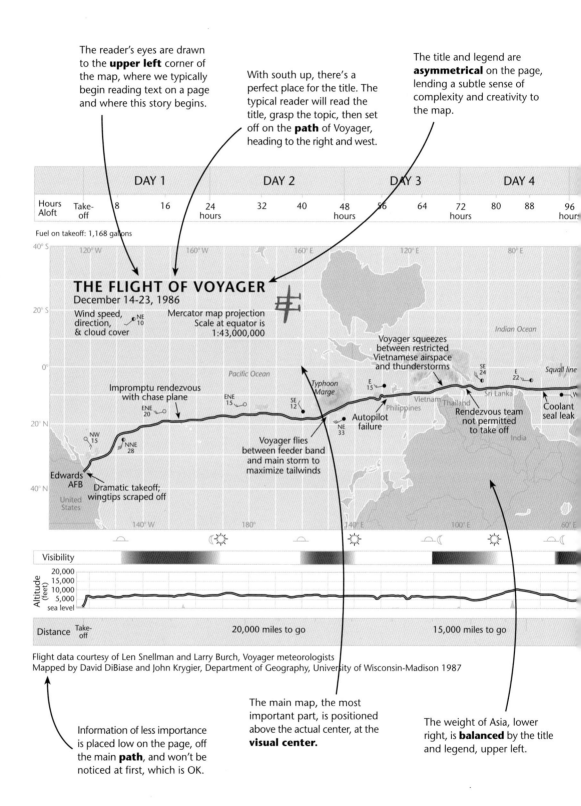

	DAY 1			DAY 2			DAY 3			DAY 4	
Hours Aloft	Take-off	8	16	24 hours	32	40	48 hours	56	64	72 hours	80	88	96 hours

Fuel on takeoff: 1,168 gallons

40° S   120° W   160° W   160° E   120° E   80° E

20° S

THE FLIGHT OF VOYAGER
December 14-23, 1986

Wind speed, direction, & cloud cover   NE 10

Mercator map projection
Scale at equator is
1:43,000,000

Indian Ocean

0°

Pacific Ocean

Voyager squeezes between restricted Vietnamese airspace and thunderstorms

SE 24   E 22   Squall line

Impromptu rendezvous with chase plane

ENE 15   SE 12

Typhoon Marge

E 15

Vietnam   Thailand   Sri Lanka

ENE 20

Philippines

Coolant seal leak

20° N   NW 15   NNE 28

Autopilot failure

NE 33

Rendezvous team not permitted to take off

India

Voyager flies between feeder band and main storm to maximize tailwinds

Edwards AFB

Dramatic takeoff; wingtips scraped off

United States

40° N

140° W   180°   140° E   100° E   60° E

Visibility

Altitude (feet)
20,000
15,000
10,000
5,000
sea level

Distance   Take-off   20,000 miles to go   15,000 miles to go

Flight data courtesy of Len Snellman and Larry Burch, Voyager meteorologists
Mapped by David DiBiase and John Krygier, Department of Geography, University of Wisconsin-Madison 1987

Information of less importance is placed low on the page, off the main **path**, and won't be noticed at first, which is OK.

The main map, the most important part, is positioned above the actual center, at the **visual center.**

The weight of Asia, lower right, is **balanced** by the title and legend, upper left.

114

The story text blocks are distributed in relation to the flight path in order to attain **balance** over the entire map.

Sight-lines on the left and right of the map are kept simple with aligned graphic elements and text.

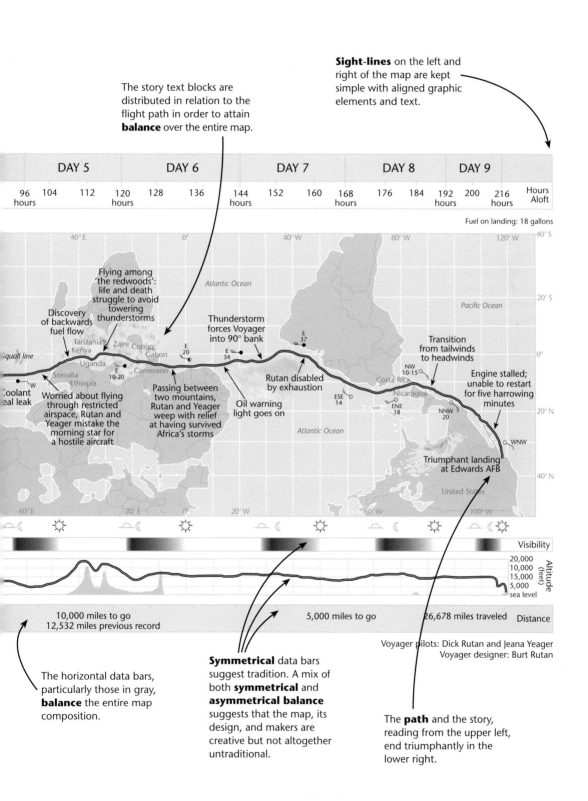

DAY 5	DAY 6	DAY 7	DAY 8	DAY 9	

| 96 hours | 104 | 112 | 120 hours | 128 | 136 | 144 hours | 152 | 160 | 168 hours | 176 | 184 | 192 hours | 200 | 216 hours | Hours Aloft |

Fuel on landing: 18 gallons

40° E   0°   40° W   80° W   120° W   40° S

20° S

Atlantic Ocean

Pacific Ocean

Flying among 'the redwoods': life and death struggle to avoid towering thunderstorms

Discovery of backwards fuel flow

Thunderstorm forces Voyager into 90° bank

Transition from tailwinds to headwinds

0°

Tanzania
Kenya
Zaire Congo
Gabon

E 20
E 34

E 37

Squall line
Uganda
Somalia
Ethiopia
E 10-20
Cameroon

NW 10-15
Costa Rica

Engine stalled; unable to restart for five harrowing minutes

Coolant seal leak
W

Worried about flying through restricted airspace, Rutan and Yeager mistake the morning star for a hostile aircraft

Passing between two mountains, Rutan and Yeager weep with relief at having survived Africa's storms

Rutan disabled by exhaustion

Oil warning light goes on

ESE 14

Nicaragua
ENE 18
NNW 20

WNW

20° N

Atlantic Ocean

Triumphant landing at Edwards AFB

40° N

United States

60° E   20° E   0°   20° W   60° W   100° W

Visibility

20,000
10,000
15,000
5,000
sea level

Altitude (feet)

10,000 miles to go
12,532 miles previous record

5,000 miles to go

26,678 miles traveled   Distance

Voyager pilots: Dick Rutan and Jeana Yeager
Voyager designer: Burt Rutan

The horizontal data bars, particularly those in gray, **balance** the entire map composition.

Symmetrical data bars suggest tradition. A mix of both **symmetrical** and **asymmetrical balance** suggests that the map, its design, and makers are creative but not altogether untraditional.

The **path** and the story, reading from the upper left, end triumphantly in the lower right.

Graphical Excellence

A different way to think about the big picture
of map design are the ideas of Edward Tufte.
Unlike the emphasis on the purely visual en-
couraged by path, center, balance, symmetry,
sight-lines, and grids, Tufte is concerned
with interesting data and complex ideas
presented with clarity and intelligence.
Below are 23 "Tufteisms" from his
books. They are intended to
help you think about your
map design choices.
They are not rules.
There are no rules.

Revise and edit.
Forgo chartjunk.
Erase non-data-ink.
To clarify, add detail.
Erase redundant data-ink.
Maximize the data-ink ratio.
Above all else, show the data.
The revelation of the complex.
Showing complexity is hard work.
Show data variation, not design variation.
Graphics must not quote data out of context.
Graphical excellence is nearly always multivariate.
Graphical excellence requires telling the truth about the data.
If the numbers are boring, then you've got the wrong numbers.
Write out explanations of the data on the graphic itself. Label important events in the data.
Graphical excellence consists of complex ideas communicated with clarity precision, and efficiency.
Clear, detailed, and thorough labeling should be used to defeat graphical distortion and ambiguity.
The number of information-carrying (variable) dimensions depicted should not exceed the number
Graphical excellence is the well-designed presentation of interesting data – a matter of substance,
In time-series displays of money, deflated and standardized units of monetary measurement are nearly
Graphical excellence is that which gives to the viewer the greatest number of ideas in the shortest
If the nature of the data suggests the shape of the graphic, follow that suggestion. Otherwise, move
The representation of numbers, as physically measured on the surface of the graphic itself, should

of dimensions in the data.
of statistics, and of design.
always better than nominal units.
time with the least ink in the smallest space.
toward horizontal graphics about 50 percent wider than tall.
be directly proportional to the numerical quantities represented.

The flight path, countries, weather, storms, key events, days, hours, fuel, visibility, altitude, distance ... and more reveal the **complex** flight of Voyager. Designing this map was really **hard work.**

Detail was added along the flight path, including the wind barbs and countries with airspace crossed by Voyager. Similar data elsewhere are not relevant to the story and thus left off the map.

Design variation reflects **data variation** on the map: vital features are included and jump out, like the flight path. Less important features, like the grid, fall back.

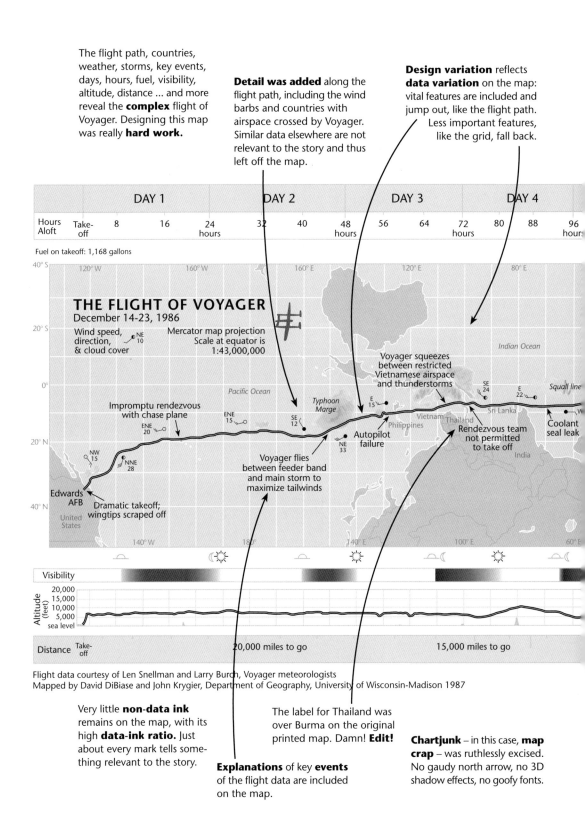

		DAY 1			DAY 2			DAY 3			DAY 4	
Hours Aloft	Take-off	8	16	24 hours	32	40	48 hours	56	64	72 hours	80	88	96 hours

Fuel on takeoff: 1,168 gallons

THE FLIGHT OF VOYAGER
December 14-23, 1986

Wind speed, direction, & cloud cover

Mercator map projection
Scale at equator is
1:43,000,000

Pacific Ocean

Indian Ocean

Voyager squeezes between restricted Vietnamese airspace and thunderstorms

Typhoon Marge

Squall line

Sri Lanka

Impromptu rendezvous with chase plane

Autopilot failure

Vietnam Thailand

Rendezvous team not permitted to take off

Coolant seal leak

India

Philippines

Voyager flies between feeder band and main storm to maximize tailwinds

Edwards AFB

Dramatic takeoff; wingtips scraped off

United States

Visibility

Altitude (feet)
20,000
15,000
10,000
5,000
sea level

Distance Take-off   20,000 miles to go   15,000 miles to go

Flight data courtesy of Len Snellman and Larry Burch, Voyager meteorologists
Mapped by David DiBiase and John Krygier, Department of Geography, University of Wisconsin-Madison 1987

Very little **non-data ink** remains on the map, with its high **data-ink ratio.** Just about every mark tells something relevant to the story.

The label for Thailand was over Burma on the original printed map. Damn! **Edit!**

Explanations of key **events** of the flight data are included on the map.

Chartjunk – in this case, **map crap** – was ruthlessly excised. No gaudy north arrow, no 3D shadow effects, no goofy fonts.

118

Details of the days, hours, fuel, visibility, altitude, and distance are shown in the five data bars spanning the map. All inform the story of Voyager.

Data are shown in **context:** geographic, meteorological, diurnal, altitudinal, experiential.

Capturing the essence of the events central to the flight in a brief sentence or phrase took repeated critical **revision**.

Scraped wingtips, typhoons, high headwinds, backwards fuel flow, exhaustion, and a mere 18 gallons of fuel left on landing are **not boring.**

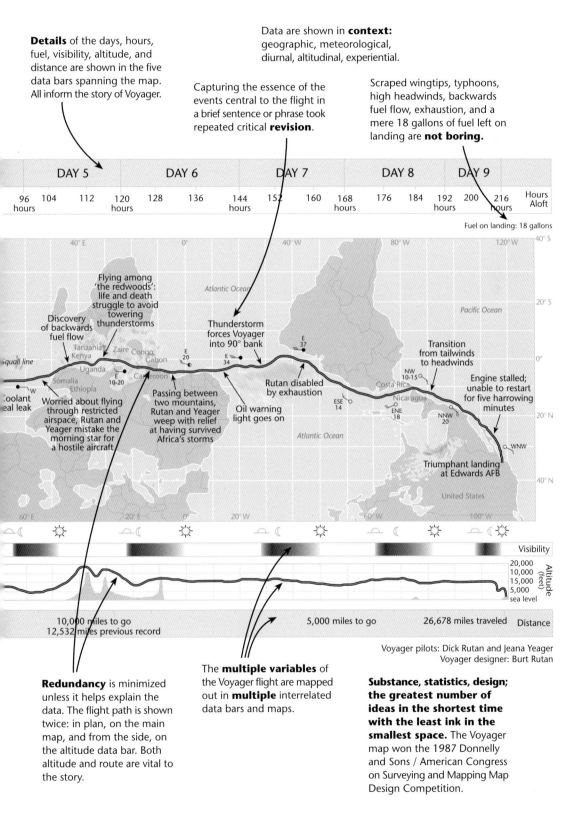

| DAY 5 | DAY 6 | DAY 7 | DAY 8 | DAY 9 |

96 hours 104 112 120 hours 128 136 144 hours 152 160 168 hours 176 184 192 hours 200 216 hours — Hours Aloft

Fuel on landing: 18 gallons

Flying among 'the redwoods': life and death struggle to avoid towering thunderstorms

Discovery of backwards fuel flow

Thunderstorm forces Voyager into 90° bank

Transition from tailwinds to headwinds

Rutan disabled by exhaustion

Engine stalled; unable to restart for five harrowing minutes

Worried about flying through restricted airspace, Rutan and Yeager mistake the morning star for a hostile aircraft.

Passing between two mountains, Rutan and Yeager weep with relief at having survived Africa's storms

Oil warning light goes on

Triumphant landing at Edwards AFB

Coolant seal leak

Squall line

Atlantic Ocean · Pacific Ocean · United States

Visibility

20,000 / 10,000 / 15,000 / 5,000 / sea level — Altitude (feet)

10,000 miles to go
12,532 miles previous record

5,000 miles to go

26,678 miles traveled — Distance

Voyager pilots: Dick Rutan and Jeana Yeager
Voyager designer: Burt Rutan

Redundancy is minimized unless it helps explain the data. The flight path is shown twice: in plan, on the main map, and from the side, on the altitude data bar. Both altitude and route are vital to the story.

The **multiple variables** of the Voyager flight are mapped out in **multiple** interrelated data bars and maps.

Substance, statistics, design; the greatest number of ideas in the shortest time with the least ink in the smallest space. The Voyager map won the 1987 Donnelly and Sons / American Congress on Surveying and Mapping Map Design Competition.

I knew every page in that atlas by heart. How many days and nights I had lingered over its old faded maps, following the blue rivers from the mountains to the sea, wondering what the little towns really looked like, and how wide were the sprawling lakes! I had a lot of fun with that atlas, traveling, in my mind, all over the world.

Hugh Lofting, *The Voyages of Dr. Doolittle* (1922)

When our maps do not fit the territory, when we act as if our inferences are factual knowledge, we prepare ourselves for a world that isn't there. If this happens often enough, the inevitable result is frustration and an ever-increasing tendency to warp the territory to fit our maps. We see what we want to see, and the more we see it, the more likely we are to reinforce this distorted perception, in the familiar circular and spiral feedback pattern.

Harry L. Weinberg, *Levels of Knowing and Existence* (1959)

The Western World has been brainwashed by Aristotle for the last 2,500 years. The unconscious, not quite articulate, belief of most Occidentals is that there is one map which adequately represents reality. By sheer good luck, every Occidental thinks he or she has the map that fits.

Robert Anton Wilson, "Robert Anton Wilson: Searching for Cosmic Intelligence" (1981)

You know how Venice looks upon the map –
Isles that the mainland hardly can let go?

Robert Browning, *Stafford* (1837)

More...

Philosophers and others are beginning to take a serious interest in information graphics. John Bender and Michael Marrinan's *The Culture of Diagram* (Stanford, 2010) comes at it from a very academic historical perspective that's nonetheless valuable. James Elkins's *How to Use Your Eyes* (Routledge, 2000) is an anthology of graphics, each of which he discusses. Almost anything by Elkins is interesting. Carolyn Handa's *Visual Rhetoric in a Digital World* (Bedford/St. Martin's, 2004) is a critical sourcebook of readings. Denis Wood and John Fels apply this kind of thinking to maps in their *The Natures of Maps* (Chicago, 2008).

Any of Edward Tufte's books on information graphics is a must-see for the big picture of map design: *The Visual Display of Quantitative Information* (Graphics Press, 1983), *Envisioning Information* (Graphics Press, 1990), *Visual Explanations: Images and Quantities, Evidence and Narrative* (Graphics Press, 1997), and *Beautiful Evidence* (Graphics Press, 2006).

Henry Beck's map of the London Underground is a design classic. The tale of its design is the subject of Ken Garland's *Mr. Beck's Underground Map* (Capital Transport, 1994). Design is not just a matter of tasteful layouts: it has profound public and political dimensions. Nate Burgos's discussion of iconic modernist designer Herbert Bayer's *World Geo-Graphic Atlas* is worth a look (www.boxesandarrows.com/view/information_ecology_bayers_book_of_maps). Janet Abrams and Peter Hall's anthology *Else/Where: Mapping* (University of Minnesota Design Institute, 2006) collects maps designed by utilizing a bewildering range of cutting-edge design perspectives.

The annual North American Cartographic Information Society (nacis.org) meeting has many events devoted to map design. For map design feedback and critique, check out the CartoTalk discussion forum (cartotalk.com).

A useful hands-on guide to page design is Timothy Samara, *Making and Breaking the Grid* (Rockport Publishers, 2005). Two sources with examples of great page design are Jaroslav Andel's *Avant-Garde Page Design* (Delano-Greenidge, 2002) and Philip Meggs's *Meggs' History of Graphic Design* (Wiley, 2005).

Sources: The Flight of Voyager map is redrawn from the original map created by David DiBiase and John Krygier (1987). The story of Voyager is documented in the book of the same name by Jeana Yeager, Dick Rutan, and Phil Patton (Knopf, 1987). *The Flight of Voyager* map was published in the *Voyager* book, split between the front and rear endpapers of the hardcover edition. George Jenks's eye movement illustrations were published in his "Visual Integration in Thematic Mapping: Fact or Fiction?" *International Yearbook of Cartography* 13, 1973 (pp. 30-31).

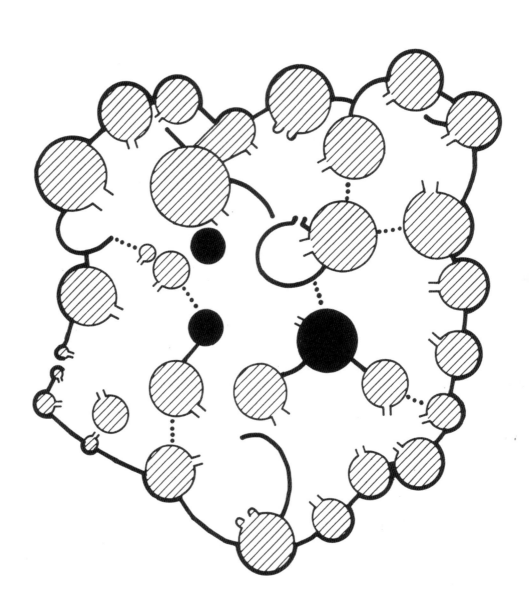

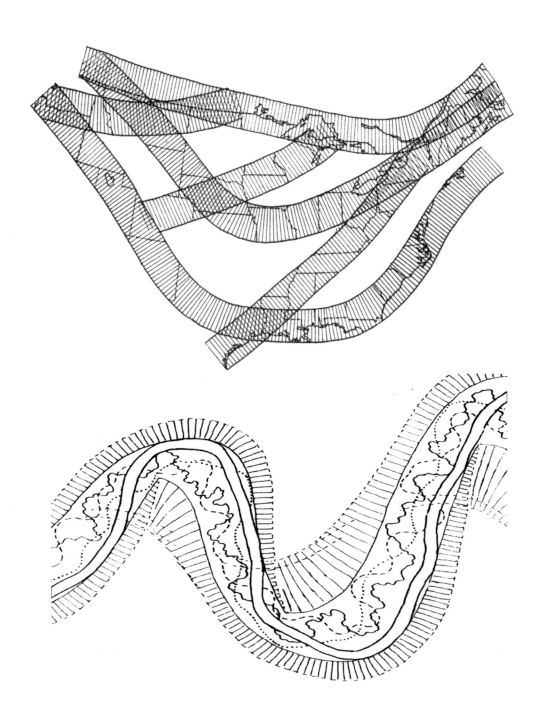

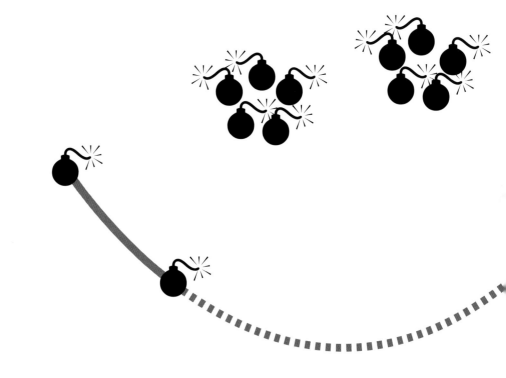

What jumps out at you?

Geo-Smiley Terror Spree

Luke Helder, a university student from Minnesota, went on a 2-week spree of bombings throughout the midwestern U.S. in an attempt to create a giant smiley face across the nation.

This is a crime that could only make sense in a map-immersed society like ours, since it is a crime that makes sense only if mapped, and then only if mapped at the the scale of the entire U.S. Zoom in, say, on the bombings that make up the eyes,

... and the pattern disappears.

The pattern jumps out only when the relevant elements are *emphasized*.

CHAPTER 7

7 The Inner Workings of Map Design

Each map piece has its own internal structure, a structure that in some ways mimics that of the map as a whole. This is especially true of the map itself, that part of the whole display that justifies calling the whole thing a map. This piece differs from the others in that the space of the map proper is geographic whereas that of the big picture is merely graphic. Here then other sorts of rhetorical strategies come into play, ways of focusing the map reader's attention that also play a role in the "big picture" but are of the essence here: figure-ground and visual difference.

Thinking about Visual Differences

Our visual lives are full of depth and contrast. Maps too can be designed with depth and contrast, enhancing their ability to effectively communicate. The perceptual effect called figure-ground is behind our ability to see visual depth. Purposeful use of figure-ground helps create effective maps. Keep in mind that figure-ground plays a meaningful role in what gets relegated to the background or left off the map entirely.

Figure-Ground

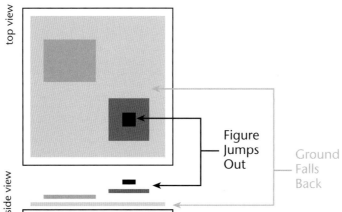

On a map, some parts stand out – figure – some fall back – ground – and others fall off the map altogether.

A successful figure-ground strategy on a map reveals what's most important first; these elements jump out. Less important elements are less visually noticeable and fall toward the back. Figure-ground relationships clearly communicate what you want your map to emphasize.

Figure on maps

    Most important map elements
    More important meaning
    Distinct form and shape
    Jumps out

Ground on maps

    Least important map elements
    Less important meaning
    Indistinct form and shape
    Falls back

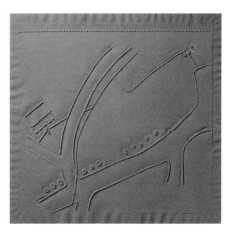

Figure-ground perception occurs with hearing, taste, smell, touch, and vision. Raised symbol maps for the blind are effective for communicating tactile figure-ground distinctions. The raised symbols literally stand out as figure over the smooth map background.

Her Figure on Our Ground

The famous Ditchley portrait of Queen Elizabeth I standing on a map of England reminds us that more is at stake in figure-ground relationships than elevating an object above the background.

On maps, the less powerful are relegated to the background or left off the map altogether, erased from our consciousness. Early maps of the Americas, Australia, and Africa communicated that there was much land available for colonization. The map silenced the voices of those already living there. It did this by burying the inhabitants below the ground.

Neither Figure Nor Ground

A typical web map reveals a system of tubes for driving around and buying stuff. Streams, wetlands, vegetation, bike trails, walking trails, and more are demoted to the background or entirely drop away. You can't show everything, but what you do show, what stands out as figure, reveals your intent for your map and your goals.

While creating your map, consider what should jump out, what should fall back, and, importantly, the implications of what falls off the map altogether.

Enhancing Visual Differences

There are diverse ways to add depth to flat maps to help the map reader see the point of your map. The abstract examples below provide ideas for establishing figure-ground relationships on your map and can be combined on your map. Not all will work on every map.

Visual difference: Noticeable visual differences separate figure from ground. To focus attention on the important areas on your map, make them visually distinct from less important areas.

Detail: Figure has more detail than ground. To focus attention on the most important area on your map, generalize and reduce detail in less important areas.

Edges: Sharp, defined edges separate figure from ground. Conversely, weak edges move less important elements on the map from figure to ground. Gray or white (reversed-out) lines and type weaken edges and move less important information to a lower visual level.

Texture: Isolated coarse textures tend to stand out as figure and move to higher visual levels.

Layering: Visual depth is enhanced when the ground appears to continue behind the figure. Grids of latitude and longitude can be visually manipulated to focus attention on the most important area of your map by appearing to run behind the most important parts.

Shape and size: Figure has shape and size. Map elements with simple closed shapes tend to be seen as figure. However, complex shapes also draw attention and tend toward figure. Larger symbols tend toward stronger figure.

130

Closure: Closed objects tend to jump out from the ground.

Proximity: Objects close together tend to stand out as figure.

Simplicity: Simple objects tend to form stronger figure.

Direction: Objects with the same orientation, heading in the same direction, tend to form figure.

Familiarity: Objects with familiar, recognizable shapes jump out as figure.

Color: Strong figure is created by intense colors, reds, and highly contrasting hues (yellow-black, white-blue). Complementary hues (red-green, blue-orange) create ambiguous figure-ground.

Visualizing Visual Differences

A design guide shows ways you can visually manipulate point, line, and area symbols on a map to achieve visual depth. Create a design guide, like the one below, based on common elements in your work. Print or view the guide in the final medium, and use it to help design your map. Subtle differences are noticeable, but if you want the map reader to see differences on your map, make sure there are substantial differences in the symbols you choose.

A Visual Design Guide

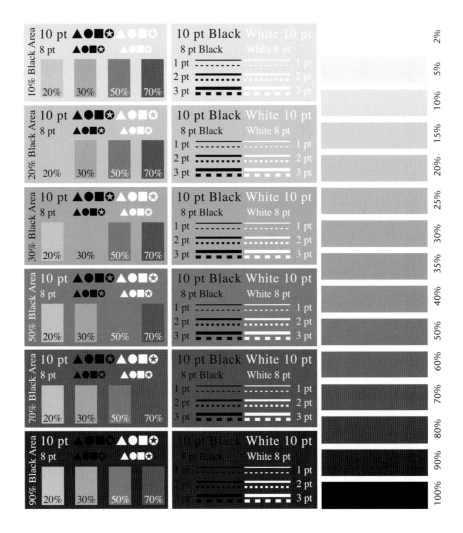

Symbol Differentiation

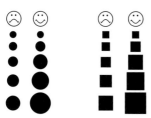  

To differentiate symbols, ensure that they are different enough to notice.

Trust your eye and err in the direction of too much difference.

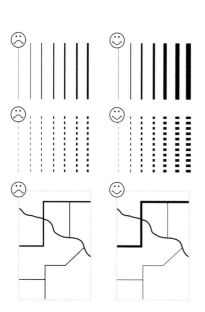

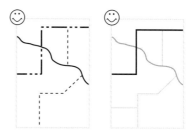

Without **visual differences** among the symbols on the map you have a nasty, unintelligible mess.

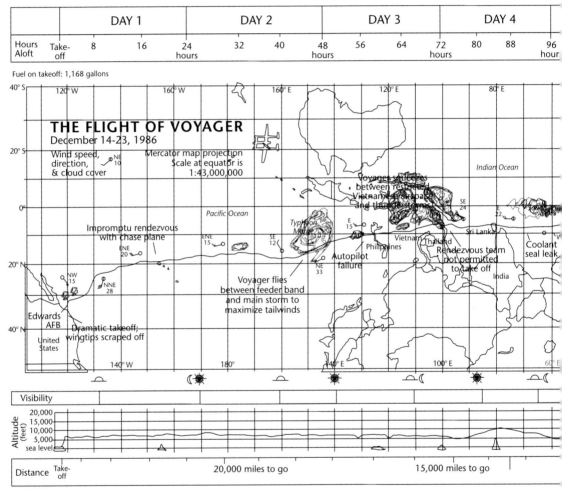

	DAY 1			DAY 2			DAY 3			DAY 4		

Hours Aloft | Take-off | 8 | 16 | 24 hours | 32 | 40 | 48 hours | 56 | 64 | 72 hours | 80 | 88 | 96 hour

Fuel on takeoff: 1,168 gallons

THE FLIGHT OF VOYAGER
December 14-23, 1986

Wind speed, direction, & cloud cover

Mercator map projection
Scale at equator is
1:43,000,000

Pacific Ocean

Indian Ocean

Impromptu rendezvous with chase plane

Typhoon Marge

Voyager squeezes between restricted Vietnam airspace and thunderstorms

Vietnam Thailand

Sri Lanka

Coolant seal leak

Autopilot failure

Philippines

Rendezvous team not permitted to take off

India

Voyager flies between feeder band and main storm to maximize tailwinds

Edwards AFB

United States

Dramatic takeoff; wingtips scraped off

Visibility

Altitude (feet)

20,000
15,000
10,000
5,000
sea level

Distance Take-off   20,000 miles to go   15,000 miles to go

Flight data courtesy of Len Snellman and Larry Burch, Voyager meteorologists
Mapped by David DiBiase and John Krygier, Department of Geography, University of Wisconsin-Madison, 1987

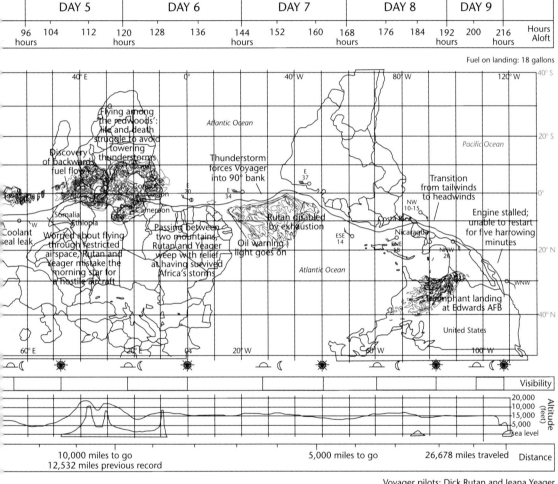

Voyager pilots: Dick Rutan and Jeana Yeager
Voyager designer: Burt Rutan

So, add **visual differences** – driven by your data and the goals for your map...

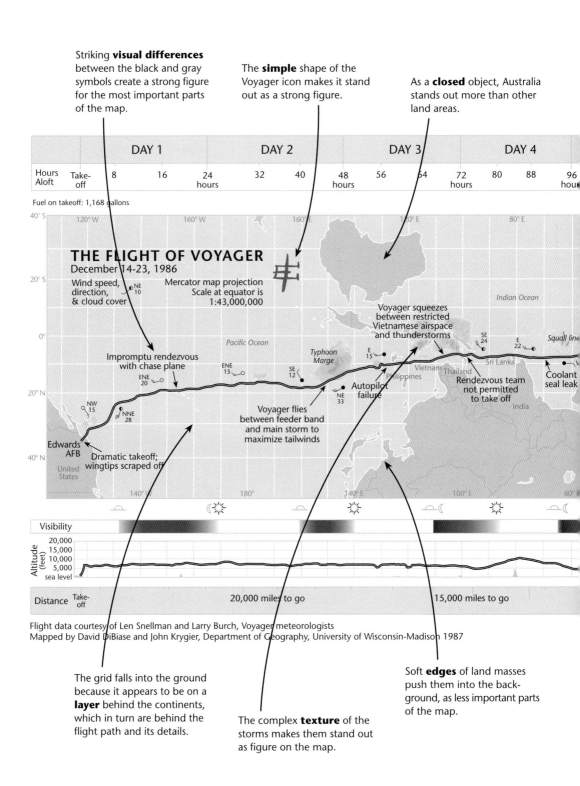

Striking **visual differences** between the black and gray symbols create a strong figure for the most important parts of the map.

The **simple** shape of the Voyager icon makes it stand out as a strong figure.

As a **closed** object, Australia stands out more than other land areas.

		DAY 1			DAY 2		DAY 3			DAY 4	
Hours Aloft	Take-off	8	16	24 hours	32	40	48 hours	56	64	72 hours	80	88	96 hour

Fuel on takeoff: 1,168 gallons

THE FLIGHT OF VOYAGER
December 14-23, 1986

Wind speed, direction, & cloud cover

Mercator map projection
Scale at equator is
1:43,000,000

Indian Ocean

Pacific Ocean

Voyager squeezes between restricted Vietnamese airspace and thunderstorms

Typhoon Marge

Squall line

Impromptu rendezvous with chase plane

Vietnam Thailand

Sri Lanka

Coolant seal leak

Autopilot failure

Philippines

Rendezvous team not permitted to take off

India

Voyager flies between feeder band and main storm to maximize tailwinds

Edwards AFB

United States

Dramatic takeoff; wingtips scraped off

Visibility

Altitude (feet)
20,000
15,000
10,000
5,000
sea level

Distance Take-off | 20,000 miles to go | 15,000 miles to go

Flight data courtesy of Len Snellman and Larry Burch, Voyager meteorologists
Mapped by David DiBiase and John Krygier, Department of Geography, University of Wisconsin-Madison 1987

The grid falls into the ground because it appears to be on a **layer** behind the continents, which in turn are behind the flight path and its details.

The complex **texture** of the storms makes them stand out as figure on the map.

Soft **edges** of land masses push them into the background, as less important parts of the map.

The east-west and north-south **directions** of the grid make it hang together as a coherent whole behind the entire map.

The distinctive **shapes** of continents make them stand out as figure; they are pushed back into the ground by the use of gray and white.

The **familiar** shape of South America creates a recognizable figure even though it is oriented south up.

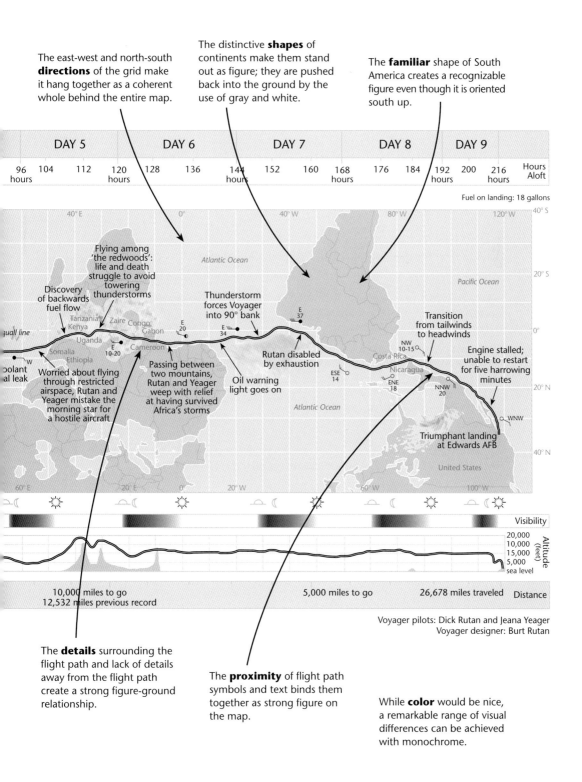

DAY 5			DAY 6			DAY 7			DAY 8		DAY 9	

96 hours	104	112	120 hours	128	136	144 hours	152	160	168 hours	176	184	192 hours	200	216 hours	Hours Aloft

Fuel on landing: 18 gallons

40° E   0°   40° W   80° W   120° W   40° S

Atlantic Ocean

20° S

Pacific Ocean

Flying among 'the redwoods': life and death struggle to avoid towering thunderstorms

Discovery of backwards fuel flow

Thunderstorm forces Voyager into 90° bank

E 37

E 20

quall line

Tanzania
Kenya   Zaire Congo
Gabon   E 20   E 34

0°

Uganda

Somalia   E 10-20   Cameroon

Transition from tailwinds to headwinds

NW 10-15

Costa Rica

ponlant   Ethiopia
al leak   W

Worried about flying through restricted airspace, Rutan and Yeager mistake the morning star for a hostile aircraft

Passing between two mountains, Rutan and Yeager weep with relief at having survived Africa's storms

Rutan disabled by exhaustion

Oil warning light goes on

ESE 14

Nicaragua
ENE 18

Engine stalled; unable to restart for five harrowing minutes

NNW 20

20° N

Atlantic Ocean

WNW

Triumphant landing at Edwards AFB

40° N

United States

60° E   20° E   0°   20° W   60° W   100° W

Visibility

20,000
10,000
15,000
5,000
sea level

Altitude (feet)

10,000 miles to go
12,532 miles previous record

5,000 miles to go

26,678 miles traveled   Distance

Voyager pilots: Dick Rutan and Jeana Yeager
Voyager designer: Burt Rutan

The **details** surrounding the flight path and lack of details away from the flight path create a strong figure-ground relationship.

The **proximity** of flight path symbols and text binds them together as strong figure on the map.

While **color** would be nice, a remarkable range of visual differences can be achieved with monochrome.

137

[looking at map] 'You are here'... Wow! How do they know? That's so cool!!

Twister, *A Shot In The Park* (2007)

He looked on the map for the Lonely Tree, but did not find it.

"That map is absolutely worthless," said Good Fortune emphatically.

"Perhaps it is only a matter of looking long enough," Christian suggested. "Yesterday I found Durben Mot, and there's a lot of sand here.... We must be here, and the place is called Gatsen Mot, and there's a well near."

"That's wonderful!" cried Good Fortune, "and those who say a man should be sparing of his words are right. Gatsen Mot is Mongolian for Lonely Tree, and I beg your pardon for my hasty speech."

Fritz Muhlenweg, *Big Tiger and Christian* (1954)

We have watched mutant creatures crawl from sewers into cold flat starlight and whisper shyly to each other, drawing maps and messages in faecal mud.

China Miéville, *Perdido Street Station* (2000)

More...

One of the best discussions of the internal structure of information graphics is Edward Tufte's in the second half of his book *The Visual Display of Quantitative Information* (1983). Another stimulating treatment is that of Richard Saul Wurman in his *Information Anxiety* (Doubleday, 1989). In cartography, see Borden Dent et al., *Cartography: Thematic Map Design* (McGraw-Hill, 2008), with its excellent overview of intellectual and visual hierarchies on maps.

On map design in general as well as the concerns of this chapter, see Cynthia Brewer, *Designing Better Maps: A Guide for GIS Users* (ESRI Press, 2005) and *Designed Maps: A Sourcebook for GIS Users* (ESRI Press, 2008).

Focused on graphs and PowerPoint but useful for map makers: Stephen Kosslyn, *Graph Design for the Eye and Mind* (Oxford, 2006) and *Clear and to the Point: Eight Psychological Principles for Compelling PowerPoint Presentations* (Oxford, 2007). But also see Tufte's screed against PowerPoint, *The Cognitive Style of PowerPoint* (Graphics Press, 2003).

Sources: The *Geo-Smiley Terror Spree* map is based on data from a map in *Time* (May 20, 2002). The tactile map is from Paul Groves and Joseph Wiedel, "The Development of Spatial Concepts Through Tactual Graphics: A Mobility and Orientation Program for Blind Children." *International Yearbook of Cartography* 13, 1973, pp. 184-188. The Ditchley portrait of Queen Elizabeth I is from the Wikimedia Commons (commons.wikimedia.org). The section of the chapter on enhancing visual differences is based on Borden Dent's chapter on intellectual and visual hierarchies in his *Cartography: Thematic Map Design* (McGraw-Hill, 2008). The symbol differentiation examples are redrawn from R.W. Anson and F.J. Ormeling (eds.), *Basic Cartography* (International Cartographic Association, 1984).

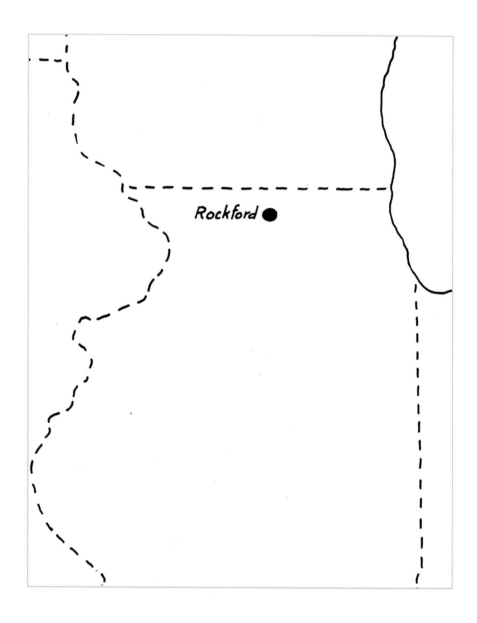

Rockford

Less is more?

Arctic

Atlantic

Amazon

Pacific

Arctic

Siberia

Gobi

Himalayas

Sahara

Indian

CHAPTER

8 Map Generalization and Classification

Fewer data can be better. Bill Bunge made a map he called *The Continents and Islands of Mankind* to make a point: as land and water barriers are about equally effective nowadays, there is no reason to keep mapping the continents and oceans when we are concerned with human affairs. This, then, is a map of places where there are more than 30 people per square mile. The shapes of the continents are obvious enough without being drawn. More importantly, they are beside the point. So leave them off the map! What matters here is where people are. Fewer data more effectively make the point.

Map Generalization

Our human and natural environments are complex and full of detail. Maps work by strategically reducing detail and grouping phenomena together. Driven by your intent, maps emphasize and enhance a few aspects of our world and deemphasize everything else. In contrast, Tufte's dictum "to clarify, add detail" serves as a critical check. Ruthlessly cut away the superfluous – "reality junk" – so you can add more levels of information to clarify and achieve your goals for the map.

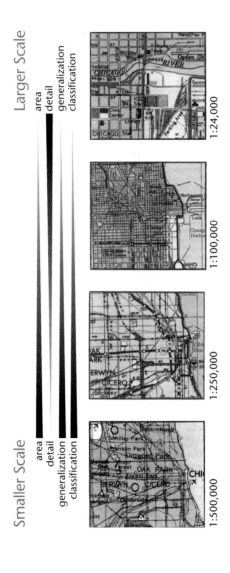

Larger Scale

area
detail
generalization
classification

Smaller Scale

area
detail
generalization
classification

1:24,000

1:100,000

1:250,000

1:500,000

Larger-Scale Maps

Less area
More detail
Less generalization
Less classification

Transformation from large to small scale requires generalization and classification.

City changes from area to point

Diversity of city sizes combined into a few categories (small, medium, large)

Minor streets and roads removed

Different types of streets and roads combined into a few categories

Houses, then major buildings, removed

Small streams removed

Detail removed from rivers and roads

Change areas on statistical maps from local to regional to national

Smaller-Scale Maps

More area
Less detail
More generalization
More classification

Selection

Maps select a few (and don't select most) features from the human or natural environment. Selected features are vital to the intent of your map. Unnecessary features should not be selected. Selection reduces clutter and enhances the reader's ability to focus on what is most important on a map.

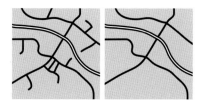

Selection is the responsibility of the map maker. Selection is automated with digital mapping. For example, at a larger scale (more detail, less area) all cities over 1000 in population are shown. At a smaller scale (less detail, more area) only cities over 100,000 population are shown.

Questions to ask when selecting map features, keeping in mind your goals for the map:

Is the feature necessary to make your point?
Will removing the feature make the map harder to understand?
If less important features are removed, do more important features stand out more clearly?
Does removal of less important features lead to a less cluttered map?

Dimension Change

A dimension change in a feature is often necessary when changing scale and useful for removing unnecessary detail from a map. A city changes from an area to a point, a river from an area to a line. Conversely, a group of points may be transformed into an area, or a group of small areas into one larger area.

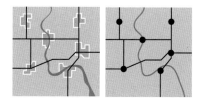

Scale change is the most common reason for dimension change, but map makers also change dimensions of features to remove clutter from a map in order to support its point. Computer mapping software can change feature dimensions based on the scale of the map.

Questions to ask that guide dimension change, keeping in mind your goals for the map:

Would changing dimensions of a feature remove unnecessary detail?
Does changing the dimensions of a feature in any way affect how it is understood by the map reader?
Does changing the dimensions of map features help make the map less cluttered?

Simplification

Features on maps are often simplified. Simplification can enhance visibility, reduce clutter, and, with digital data, reduce the size of the digital map file. Smaller-scale maps (showing more area) tend to have more simplified features than larger-scale maps. Eliminate detail that is not necessary for the map to make your point.

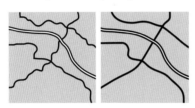

Simplification of features is done to the point where they are less complex, yet still recognizable. Identifying characteristics should be retained. Mapping software simplifies features by methodically removing detail. A line is simplified by removing every other point, for example.

Questions to ask when simplifying features, keeping in mind your goals for the map:

How simplified can a feature be and still be recognized?
Does the removal of detail remove any vital information?
Does the simplification of a feature make it more noticeable?
Does the simplification of a feature make the map less cluttered looking?

Smoothing

Smoothing map features reduces their angularity. Smoothing is related to simplification but focuses on adjustments in the location of a feature or possibly the addition of detail. Smoothing affects the qualitative look of features.

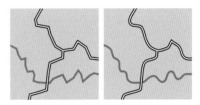

Map features that are naturally sinuous (often natural features) are more heavily smoothed. Features that are not smooth in the real world are lightly smoothed or not smoothed at all. Computers smooth features by rounding angles that exceed a set limit.

Questions to ask when smoothing features, keeping in mind your goals for the map:

How much can you smooth a feature without losing its character?
Does smoothing a feature make it more difficult to recognize?
Does smoothing a feature make it easier to recognize?
Does the smoothing of a feature make the map less cluttered looking?

Displacement

Displacement moves vital map features, that visually interfere with one another apart. Moving features away from their actual location may sound like a terrible idea, but it makes the features easier to distinguish and understand.

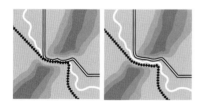

Scale change is a common reason for displacement. Displacement of map features sacrifices location accuracy for visual clarity. Displacement of point and line features is common on maps where such features are crowded together in certain areas.

Questions to ask that guide displacement, keeping in mind your goals for the map:

Are important map features interfering with one another?
Will the slight movement of a map feature make it and neighboring features easier to distinguish?
Will the slight movement of a map feature lead to confusion because the feature has been moved?

Enhancement

Enhancement of map features occurs when the map maker knows enough about the feature being mapped to add details that aid in understanding. A few bumps, for example, are added to a road the map maker knows is winding and angular so that the map reader understands the actual character of the road.

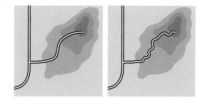

Enhancement adds detail, as opposed to removing detail, which most generalization procedures do. Enhancement must be used with care. The enhancement should not be deceptive, but instead should help the map reader understand important features on the map.

Questions to ask that guide enhancement, keeping in mind your goals for the map:

Do you know enough about a feature to enhance it?
Will enhancement help the map reader to better understand the feature and the map?
Does enhancing a feature make it easier to recognize?
Could enhancing a feature possibly lead to misunderstanding by the map reader?

For the map, the entire book-length story of the flight of Voyager was **simplified** into 18 briefly described episodes capturing the essence of the story.

In a few places, near storms, the flight path was **enhanced,** showing more detail when supported by the pilot's account (and not captured by the cruder locational data of the flight path).

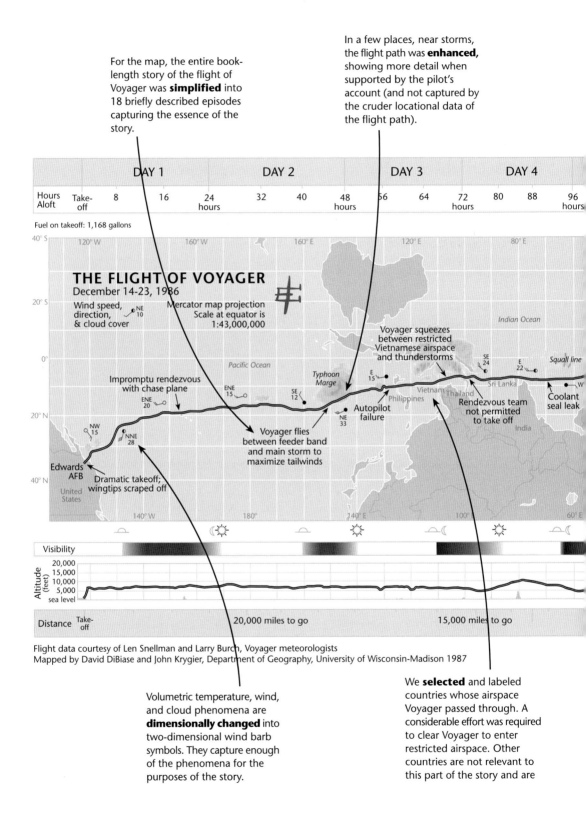

| | DAY 1 | | | | DAY 2 | | | DAY 3 | | | DAY 4 | |

| Hours Aloft | Take-off | 8 | 16 | 24 hours | 32 | 40 | 48 hours | 56 | 64 | 72 hours | 80 | 88 | 96 hours |

Fuel on takeoff: 1,168 gallons

40° S   120° W   160° W   160° E   120° E   80° E

THE FLIGHT OF VOYAGER
December 14-23, 1986

20° S

Wind speed, direction, & cloud cover
NE 10

Mercator map projection
Scale at equator is 1:43,000,000

Indian Ocean

Voyager squeezes between restricted Vietnamese airspace and thunderstorms

SE 24   E 22   *Squall line*

0°   *Pacific Ocean*   *Typhoon Marge*   E 15   *Sri Lanka*   W

Impromptu rendezvous with chase plane   ENE 15   SE 12   *Vietnam* *Thailand*   Coolant seal leak

ENE 20   *Philippines*   Rendezvous team not permitted to take off

20° N   NW 15   NNE 28   Voyager flies between feeder band and main storm to maximize tailwinds   NE 33   Autopilot failure   *India*

Edwards AFB
Dramatic takeoff; wingtips scraped off

40° N   *United States*

140° W   180°   140° E   100°   60° E

Visibility

Altitude (feet)
20,000
15,000
10,000
5,000
sea level

Distance   Take-off   20,000 miles to go   15,000 miles to go

Flight data courtesy of Len Snellman and Larry Burch, Voyager meteorologists
Mapped by David DiBiase and John Krygier, Department of Geography, University of Wisconsin-Madison 1987

Volumetric temperature, wind, and cloud phenomena are **dimensionally changed** into two-dimensional wind barb symbols. They capture enough of the phenomena for the purposes of the story.

We **selected** and labeled countries whose airspace Voyager passed through. A considerable effort was required to clear Voyager to enter restricted airspace. Other countries are not relevant to this part of the story and are

Coastlines are **simplified** by reducing the number of points in the line. Details of the coastlines are not necessary for the story of Voyager.

Coastlines are **smoothed**, reducing overall angularity. Coastlines are natural features that typically don't form sharp angles.

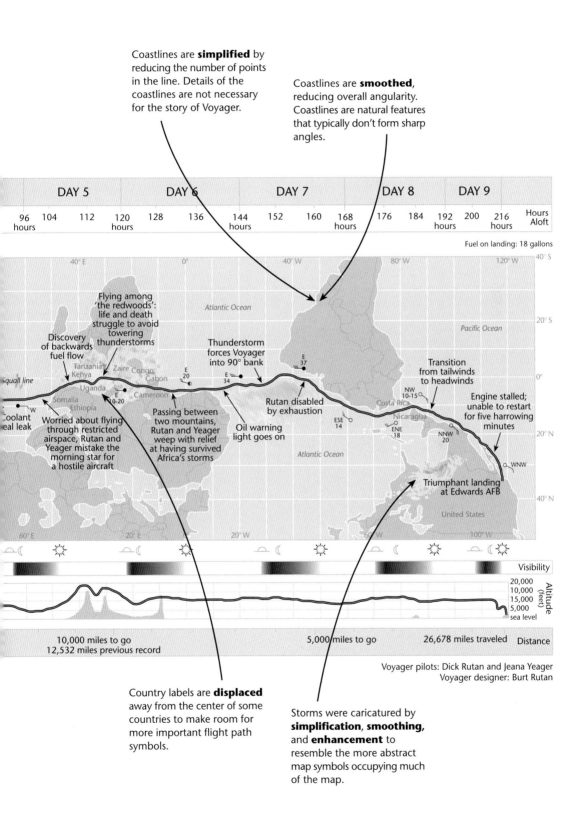

DAY 5	DAY 6	DAY 7	DAY 8	DAY 9	

96 hours · 104 · 112 · 120 hours · 128 · 136 · 144 hours · 152 · 160 · 168 hours · 176 · 184 · 192 hours · 200 · 216 hours · Hours Aloft

Fuel on landing: 18 gallons

40° E · 0° · 40° W · 80° W · 120° W · 40° S

Atlantic Ocean

Pacific Ocean

20° S

Flying among 'the redwoods': life and death struggle to avoid towering thunderstorms

Discovery of backwards fuel flow

Thunderstorm forces Voyager into 90° bank

Transition from tailwinds to headwinds

E 37

0°

Squall line

Tanzania Kenya · Zaire Congo · Gabon

Uganda · Cameroon

Somalia · Ethiopia

Coolant seal leak

W

Worried about flying through restricted airspace, Rutan and Yeager mistake the morning star for a hostile aircraft

Passing between two mountains, Rutan and Yeager weep with relief at having survived Africa's storms

Rutan disabled by exhaustion

Oil warning light goes on

E 20

E 34

E 10-20

Costa Rica

NW 10-15

Engine stalled; unable to restart for five harrowing minutes

Nicaragua

ESE 14

ENE 18

NNW 20

WNW

20° N

Atlantic Ocean

Triumphant landing at Edwards AFB

40° N

United States

60° E · 20° E · 20° W · 60° W · 100° W

Visibility

20,000 · 10,000 · 15,000 · 5,000 · sea level · Altitude (feet)

10,000 miles to go
12,532 miles previous record

5,000 miles to go

26,678 miles traveled · Distance

Voyager pilots: Dick Rutan and Jeana Yeager
Voyager designer: Burt Rutan

Country labels are **displaced** away from the center of some countries to make room for more important flight path symbols.

Storms were caricatured by **simplification**, **smoothing**, and **enhancement** to resemble the more abstract map symbols occupying much of the map.

Data Classification

Every map is generalized. It doesn't matter if it's a detailed topographic survey sheet showing trees, buildings, etc., or a map of populations. Classification operates in every map as well, distinguishing trees from buildings on the topographic map and breaking populations into classes such as small, medium, and large. Data classification is shaped by your goals for the map. In general, features in the same class should be more similar than dissimilar; features in different classes should be more dissimilar than similar.

Qualitative Point Data

A student polls community members on social issues. The first, unclassified, map is not very revealing. The classification on the second map is OK, but the third reveals more about the political landscape. Include the unclassified data so map viewers can decide if your classification is justified.

Qualitative Line Data

Roads are often classified in terms of who builds and maintains them (federal, state, local). However, this classification is not the best if your map is for tourists. Your goal for the map (tourism) should shape your data classification. Choose tourism-based classes for the roads.

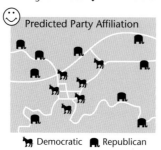

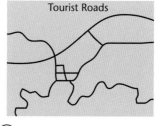

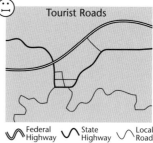

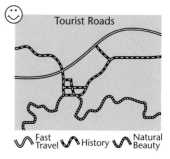

Qualitative Area Data

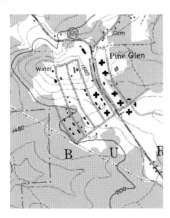

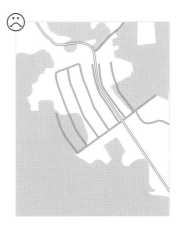

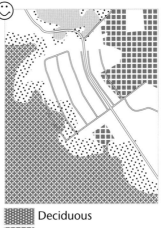

Be aware that qualitative data on maps has been classified. Determine the criteria for classification. A map may classify data based upon criteria suitable for one purpose but not necessarily for others.

For ecology projects, U.S. Geological Survey (USGS) topographic maps (above left and right) are often consulted. They classify vegetation into two categories: vegetation (gray areas) or no vegetation. One might assume that the classification is based on ecological criteria.

Alas, vegetation areas are actually classified based on military criteria. Vegetation areas are those with tree cover at least 6 feet tall that can hide military troops. It is a classification for army guys, not ecologists!

Instead, find a vegetation map (right) with a classification based on ecological criteria.

Deciduous
Coniferous
Mixed scrub

Quantitative Point Data

A map created for a community meeting about well test results should clearly show what is most important: whether the well water is safe for humans or not. The third map, below, is best for that goal.

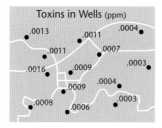

Toxins in Wells (ppm)

.0013 .0011 .0004
.0011 .0007
.0016 .0009 .0003
.0009 .0004
.0008 .0006 .0003

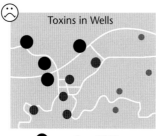

Toxins in Wells

● more than .0010 ppm
◐ .0006 - .0009 ppm
• less than .0005 ppm

Toxins in Wells

Exceeds federal toxin limits
Action: close well

Safe for non-human use

Safe for all

Quantitative Line Data

A map intended to help guide the restructuring of police patrol routes should classify data, in this case average vehicle speeds, in categories appropriate to the task: increase, maintain, or decrease patrols.

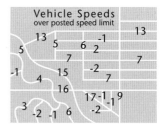

Vehicle Speeds
over posted speed limit

13 13
5 -1
5 6 2
7
-1 15 -2 7
4 16
17 -1 9
3 -2 -1 6 -1
-2

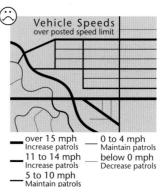

Vehicle Speeds
over posted speed limit

— over 15 mph
Increase patrols
— 11 to 14 mph
Increase patrols
— 5 to 10 mph
Maintain patrols

— 0 to 4 mph
Maintain patrols
— below 0 mph
Decrease patrols

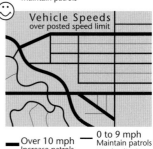

Vehicle Speeds
over posted speed limit

— Over 10 mph
Increase patrols

— 0 to 9 mph
Maintain patrols
— Under limit
Decrease patrols

154

Quantitative Area Data

With quantitative data aggregated in areas, first decide the number of classes. Fewer classes often result in distinct patterns; more classes often result in complex patterns. Which option is best depends on why you are making the map. This map shows the density of mobile homes (dark equals higher density).

In addition to choosing the number of classes, you must decide where to place boundaries between the classes. Classification schemes set these boundaries. The maps below show the density of mobile homes (dark equals higher density) unclassified and as a 5-class map using four different classification schemes.

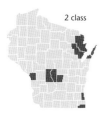
2 class

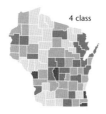
4 class

8 class

Unclassified

Unclassified

Equal Interval

Quantile

Natural Breaks

Unique

A 2-class map is good for binary (yes|no) data or data with negative and positive values

4 to 8 classes ensures that typical map readers can see distinct patterns and match a particular shading on the map to the legend

Over 8 classes produces more complex patterns, but map readers may not be able to match a shading on the map to the legend

Unclassified data (each area has a unique shading corresponding to its unique value) produce the most complex patterns

Reclassify your data with different numbers of classes and look at how the patterns change. Think about your data and goals for the map, and make an intelligent decision

Thinking Drives Classification

Poverty is a contentious issue. Debates rage over defining poverty, why it exists, and how to address it. The U.S. Census Bureau provides official data on poverty in the U.S., and various classifications of Census 2000 poverty data follow.

It is easy to calculate the percent of people in each county in the U.S. who live in a state of official poverty. But choosing how to map the data is not as easy. Common (and equally valid) data classification schemes – methods for placing boundaries between the classes on a map – are easy to generate but difficult to choose from. Understanding the benefits and problems with each classification scheme is vital, as is clarifying why you are making the map. Together, these guide the thinking behind choosing the most appropriate classification scheme for your data.

Graphing Data

Selecting a classification scheme without examining your data as a graph is a bad idea. As examples in this section reveal, classification schemes can mask important characteristics of your data and even undermine the goal of your map.

Decent mapping software will generate a histogram for you while you are classifying your data. Or, make your own. The x-axis is your data variable (from low to high) and the y-axis the number of occurrences of each value.

The poverty data have a cluster of counties near the lower to mid-end of the graph, with a smaller number of counties skewed out to 57%. On a histogram you can see where a classification scheme places class boundaries, which values are grouped together, and which values are in different groups.

If a particular classification scheme seems to violate the basic classification rule (features in the same class should be more similar than dissimilar; features in different classes should be more dissimilar than similar), then consider a different scheme. Consider placing the graph on your final map, so map users can see how the data are classified.

Unclassified Scheme

To create an unclassified scheme assign a unique visual shade to every unique data value. In essence, each unique data value is in its own class. Unclassified schemes make complex and subtle patterns by minimizing the amount of generalization.

This map, due in large part to the concentration of counties near the low end of the range of values, suggests that poverty is not a significant issue in most places, that the number of people living in poverty is somewhat similar across the U.S., and that there are few places with very high poverty.

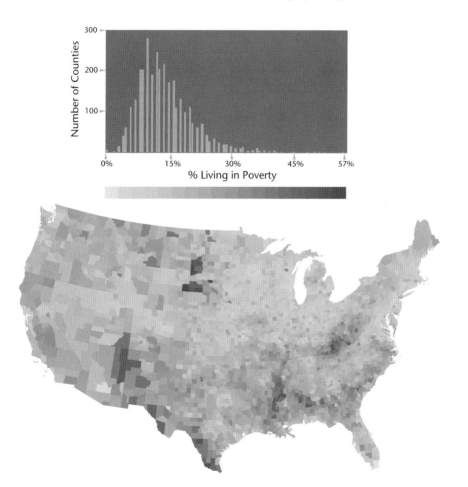

Quantile Scheme

Quantile schemes place the same number of data values in each class. Quantile schemes are appealing in that they always produce distinct map patterns. A quantile classification will never have empty classes or classes with only a few or too many values. Quantile schemes look great.

The problem with quantile schemes is that they often place similar values in different classes or very different values in the same class. The map suggests that poverty is a significant issue in many counties, and the numerous counties in the top (darkest) classes impart a rather ominous view of poverty in the United States.

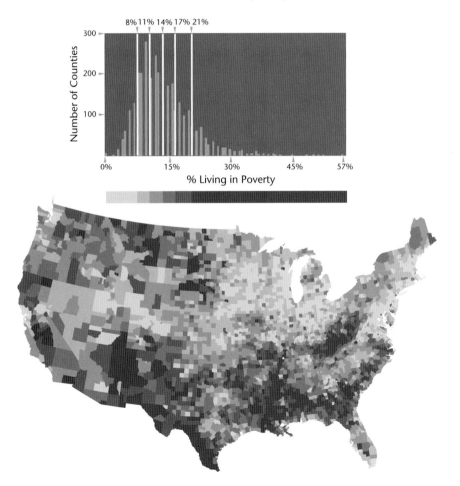

Equal-Interval Scheme

Equal-interval schemes place boundaries between classes at regular (equal) intervals. Equal-interval schemes are easily interpreted by map readers and are particularly useful for comparing a series of maps (which necessitates a common classification scheme).

Equal-interval schemes do not account for data distribution, and may result in most data values falling into one or two classes, or classes with no values. The map suggests that poverty is not an issue in most places, as there are relatively few counties in the highest three classes.

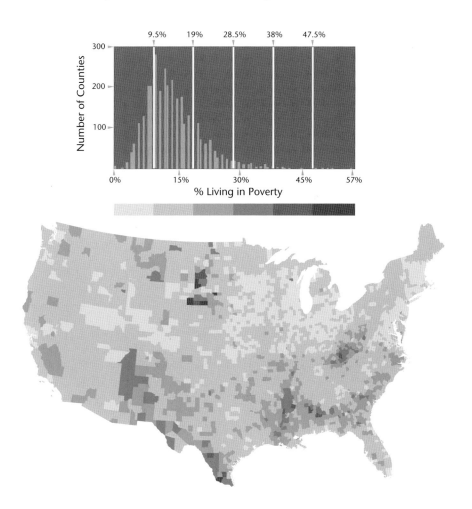

Natural-Breaks Scheme

Natural-breaks schemes minimize differences between values within classes and maximize differences between values in different classes. Class boundaries are determined by algorithms in mapping software that seek statistically significant groupings in a set of data.

Natural-breaks schemes can serve as a default classification scheme, as they take into account characteristics of the data distribution. This map makes poverty seem more significant than the equal-interval map does, but it is not quite as ominous as the quantile scheme.

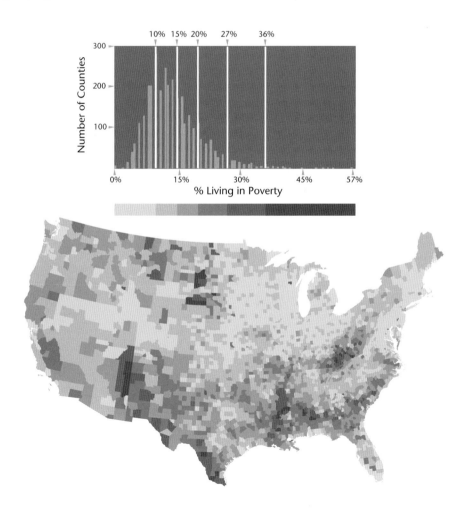

Unique Scheme

Class boundaries can be set by external criteria. A government program offers special funds to counties with over 25% poverty. A two-class map shows which counties qualify (and which don't). The map below is classified for a study of counties with very high poverty.

The researchers are not interested in counties with less than a 25% poverty rate. The remaining data are divided into classes that will aid in the analysis of high-poverty counties. Excellent for the study, this scheme is not good in general, as it suggests poverty is isolated and rare.

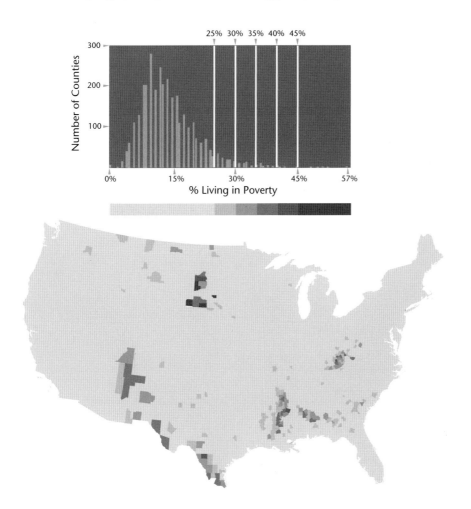

Think!

Looking at a number of classification schemes brings forth geographic facts, both the facts that are variously emphasized and those that are preserved through every variation.

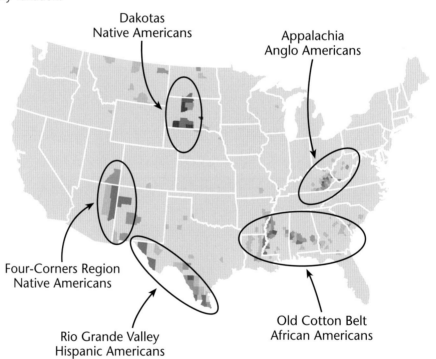

Dakotas
Native Americans

Appalachia
Anglo Americans

Four-Corners Region
Native Americans

Rio Grande Valley
Hispanic Americans

Old Cotton Belt
African Americans

The unique scheme displays counties with the highest rates of poverty. While this pattern can be seen in every scheme, the unique scheme isolates, and so draws attention to, regions of the country where poverty is a long-standing reality. The unique scheme picked out the regions of significant social injustice.

Unless you looked at a lot of maps, you might not have identified these regions of injustice as anything other than those with high levels of poverty.

It takes many different maps to begin to make sense of the world.

162

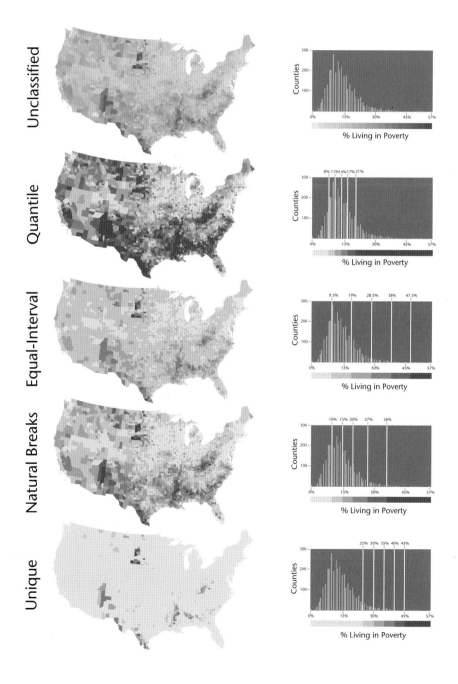

For five million pounds, I'd want a map that showed me looking at the map I'd just bought.

Rich Hall, *QI* (2005)

Steve: [completely doped on the nitrous oxide] But, this map is heavy ... it's got all of
   those ... robes on it. Robes? Rogues?
Mike: [also stoned; giggling] Roads! Steve, Aahahahaha!
Mike: [suddenly stops laughing] I'm stoned ... so are you!

Black Sheep (1996)

Airey's Railway Map is almost unique in its way, devoting itself to its subject with a singleness
of purpose which is really almost sublime, and absolutely ignoring all such minor features
of the country it portrays as hills, roads, streets, churches, public buildings, and so forth. It
is rather startling at first to find the Metropolitan Railway pursuing its course through a
country as absolutely devoid of feature as was the "Great Sahara" in the good old African
maps...

Charles Dickens, *London Guide* (1879)

I have witnessed the massacre
I am a victim of the map.

Mahmoud Darwish, *I Have Witnessed the Massacre* (1977)

164

More...

Mark Monmonier includes a succinct discussion of map generalization in *How to Lie with Maps* (University of Chicago Press, 1996). A comprehensive source for data classification is Terry Slocum et al. *Thematic Cartography and Geovisualization* (Prentice Hall, 2008). The section "Visual and Statistical Thinking: Displays of Evidence for Making Decisions" from Edward Tufte's *Visual Explanations* (Graphics Press, 1997) dramatizes the power of thinking visually about statistics.

For very deep thoughts about classification in general, see Geoffrey Bowker and Susan Leigh Star, *Sorting Things Out: Classification and Its Consequences* (MIT Press, 1999).

A curious tale about what happens when you *don't* generalize is Jorge Luis Borges's short story "Funes the Memorius" (in *Labyrinths: Selected Stories and Other Writings,* New Directions, 1964).

Sources: Bill Bunge's "Continents and Islands of Mankind" is re-created from his *Field Notes: Discussion Paper No. 2* (self published, no date). Much of this chapter is based on Borden Dent et al., *Cartography: Thematic Map Design* (McGraw-Hill, 2008), and Terry Slocum et al., *Thematic Cartography and Geovisualization* (Prentice Hall, 2008).

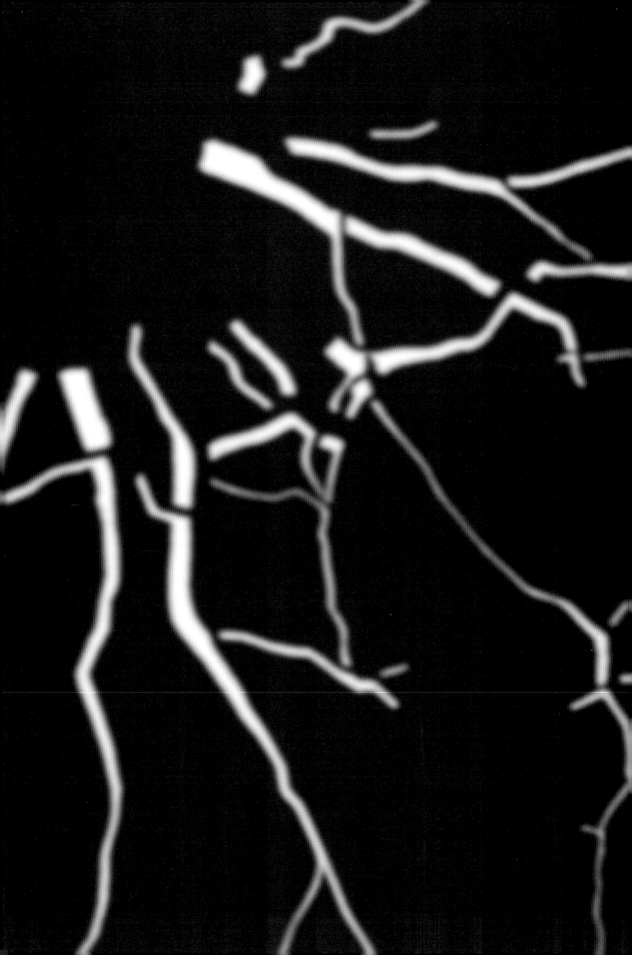

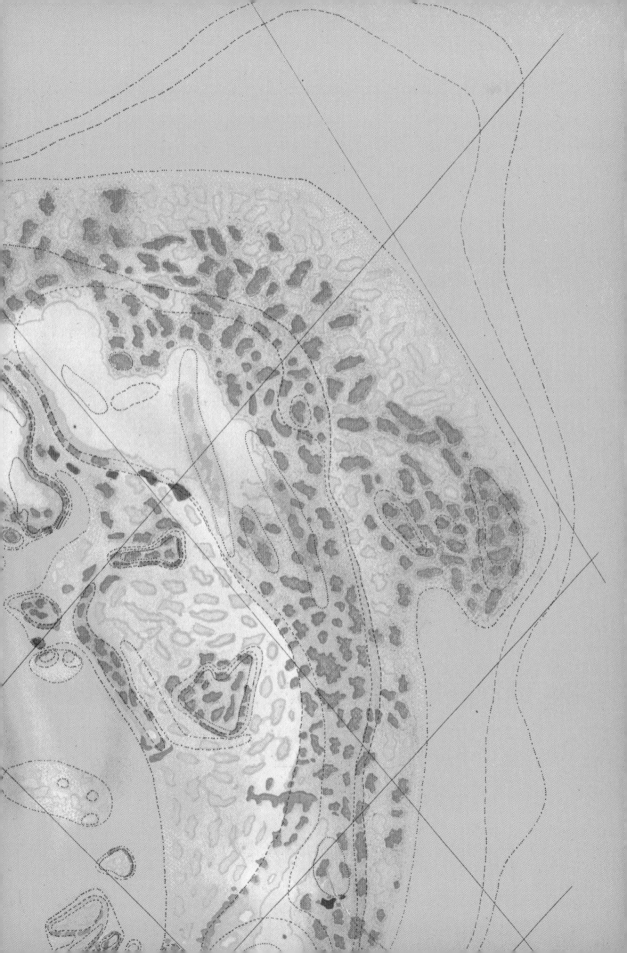

Cave

Small Muslim
Shrine or Tomb

Patrol
Route

Gantry
Crane

Manholes along
Sewer line

Thorny
Scrub

Pigeon
Roost

Oil
Seepage

Demolished
Building

Forest Ranger's
House

Stone
Mosque

Train Roundhouse
and Track

Fishing
Stakes

Shipwreck

Fenced
Fuel Tanks

Dry
Dock

Essential
Oil Plants

Semaphore
Signal

Fir Tree
Landmark

Tank
Trap

Common
Grave

Fog
Station

Whirlpool

Trigonometrical
Point on
Burial Mound

Gas Drip
Manhole

Corduroy
Road

Cairn

Stone or Brick
Wall

Ruins

Bunker

Log
Chute

Railroad
Ferry

Moss and
Grass

Pine or Cedar
Landmark

Birch
Grove

Winter
Road

Casuarina

Marabout

Cattle
Run

Wooden
Windmill

Overland Oil
Pipe & Station

Prospect
Well

Post Relay
Center

Radio
Station

Forest Service
Station

Kilometer
Post

Tent

Narrow Gauge
Railroad

Oil
Derrick

Windfall

Flour
Mill

Rail or
Wattle Fence

Trench,
Communication Trench,
& Dugout

Driftwood
Accumulation

Pit

Rail
Lubricators

Larch Tree
Landmark

Takyrs
(salt clay flats)

Volcanic
Manifestation

Polar
Station

Stone
Church

Forest
Boundary

Hedge

Slipway

Explosives
Area

Cotton

Dismantled
Railroad

Search-
light

Fireproof
Building

Artificial
Embankment

Saw
Mill

Power
Station

Fire Hydrants
along water line

Burial
Mound

Cemetery
with Trees

Metal
Fence

Vegetable
Storage

Fenced
Apiary

Statue or
Monument

Emergency
Landing Field

Cattle Burying
Ground

Patrol
Look-out

Quarantine

Leased Land
Boundary

Nine

How do you make data into visual marks?

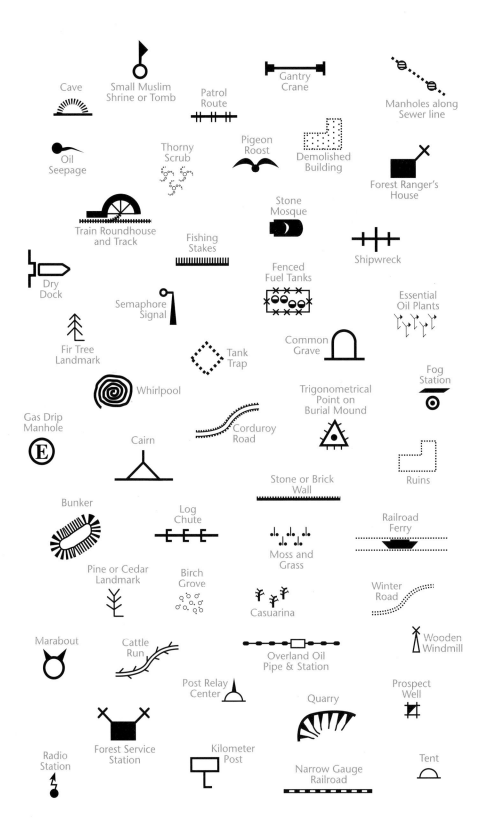

Cave

Small Muslim Shrine or Tomb

Patrol Route

Gantry Crane

Manholes along Sewer line

Oil Seepage

Thorny Scrub

Pigeon Roost

Demolished Building

Forest Ranger's House

Train Roundhouse and Track

Fishing Stakes

Stone Mosque

Shipwreck

Dry Dock

Semaphore Signal

Fenced Fuel Tanks

Essential Oil Plants

Fir Tree Landmark

Tank Trap

Common Grave

Fog Station

Whirlpool

Trigonometrical Point on Burial Mound

Gas Drip Manhole

Cairn

Corduroy Road

Ruins

Bunker

Log Chute

Stone or Brick Wall

Railroad Ferry

Pine or Cedar Landmark

Birch Grove

Moss and Grass

Winter Road

Casuarina

Marabout

Cattle Run

Overland Oil Pipe & Station

Wooden Windmill

Post Relay Center

Quarry

Prospect Well

Radio Station

Forest Service Station

Kilometer Post

Narrow Gauge Railroad

Tent

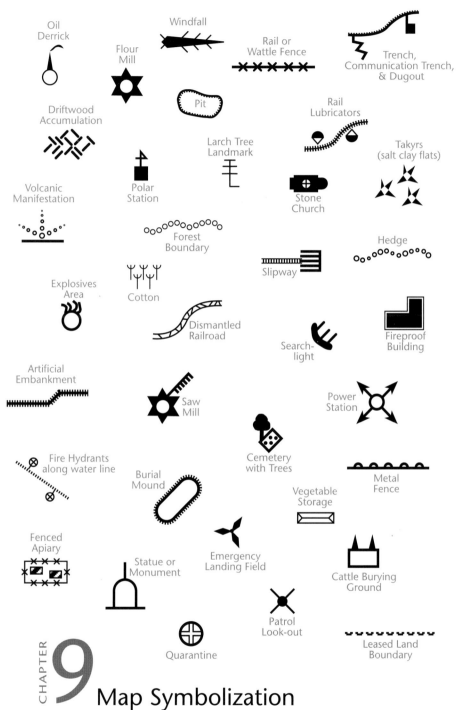

Oil Derrick

Windfall

Flour Mill

Rail or Wattle Fence

Trench, Communication Trench, & Dugout

Pit

Driftwood Accumulation

Rail Lubricators

Larch Tree Landmark

Takyrs (salt clay flats)

Volcanic Manifestation

Polar Station

Stone Church

Forest Boundary

Hedge

Explosives Area

Cotton

Slipway

Dismantled Railroad

Search-light

Fireproof Building

Artificial Embankment

Saw Mill

Power Station

Fire Hydrants along water line

Burial Mound

Cemetery with Trees

Metal Fence

Vegetable Storage

Fenced Apiary

Statue or Monument

Emergency Landing Field

Cattle Burying Ground

Patrol Look-out

Leased Land Boundary

Quarantine

9 Map Symbolization

Map symbols allow us to put just about anything on a map. These symbols are from many old sources, including 1950s Soviet topographic maps, the U.S. Coast and Geodetic Survey, the Federal Board of Surveys and Maps, U.S. Forest Service maps, Cleveland Regional Underground Survey maps, the American Railway Engineering Association, and the former U.S. War Department.

Ways to Think about Map Symbols

Everything on a map is a symbol. Map symbols, or signs, have two parts. The first is conceptual: an earthquake epicenter, a cold front, a sphere of influence. The second is a graphic mark. The mark is connected to the concept by a code or convention. For example, a cold front is often, though not always, shown as a blue line with regularly spaced triangles pointing in the direction of the front's movement:

Resemblance

Some map symbols look like particular data or concepts. A map showing the location of airports uses an airplane symbol. Airplanes make us think of airports.

Maps in a war atlas use red explosion symbols to show the location of battles. The symbol looks like an explosion, and we think of danger or conflict.

Relationship

Some map symbols intuitively suggest general kinds of data. A map showing the population of different cities uses circle sizes from small to large: sizes vary in amount, as do the data.

A map showing restaurants, antique stores, and museums in a town uses different shapes; shapes vary in kind, as do the data.

Convention

Of course, all map symbols are symbols by convention. But this is particularly clear when symbols reveal cultural bias or don't resemble what they symbolize. The U.S. Geological Survey uses a Christian cross to symbolize all places of worship – church, mosque, synagogue. Fail!

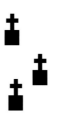

Most maps use blue for water. But water is not usually blue. Except on maps. It's a convention. If you depart from conventions (color water its actual color) you may confuse your map's readers.

Difference

All symbols work by being different from other symbols. But some symbols can be developed from others by using a process of visual differentiation.

══════ Interstate Highway	═ ═ ═ ═ Under Construction
────── State Highway	─ ─ ─ ─ Under Construction
────── County Highway	─ ─ ─ ─ Under Construction
- - - - - Other Road	Through Road in Populated Place

Standardization

Isotype consists of a series of "universally communicable" symbols. Such standards aim to reduce ambiguity through a shared set of common map symbols.

Unconvention

Old maps reveal startling, unconventional map symbols, often conventions of the past. This 17th-century Russian map contains very unconventional symbols for trees, rivers, and properties.

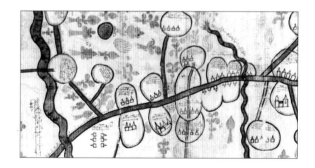

Symbols Are Graphic Marks Tied to Concepts

Wherever the graphic mark goes, the concept has to follow. Different graphic marks shift the concept, as with "tree" from the thoroughly generalized to the wholly individual. What kind of "tree" do you want? You decide. It's your map.

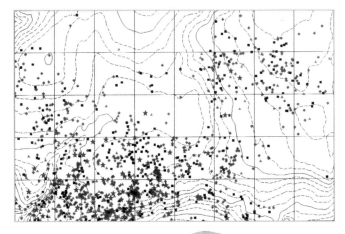

In this census plot, *Prioria copaifera* is distinguished solely by size, with adults shown as stars. This is one way to think about trees.

Tree symbols from 18th-century Russian maps are more individualized and differentiate deciduous trees from conifers.

A wholly individualized "symbol" of an actual tree from the parish map of Highbrook, in West Sussex, England. The map was made to celebrate both the millennium and the uniqueness of Highbrook village.

Symbols Are Concepts Tied to Graphic Marks

Wherever the concept goes, the graphic mark has to follow. Different concepts shift the graphic mark, as with maps of the distribution of $700 billion in corporate welfare ("Troubled Asset Relief Program"). Public interest in the TARP funds the U.S. government pumped into banks produced a variety of maps showing where the money went. The differences among them turn on their different concepts of "bank".

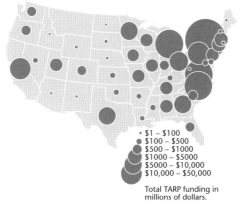

$1 – $100
$100 – $500
$500 – $1000
$1000 – $5000
$5000 – $10,000
$10,000 – $50,000

Total TARP funding in millions of dollars.

A map created by Richard Florida for *The Atlantic* conceptualizes banks as points located at the addresses of their headquarters. Because bank headquarters are so heavily concentrated in financial centers, it looks as though the TARP funds ended up in the eastern U.S. The question this map answers is:

"Where are bank headquarters located?"

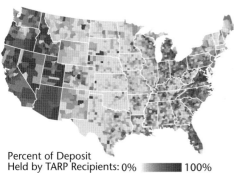

Percent of Deposit
Held by TARP Recipients: 0% ▬▬▬ 100%

A map created by Kevin Webb for Subsidyscope, conceives of banks as areally diffuse webs of activity. The Bank of America is headquartered in Charlotte, North Carolina (and hence the huge circle over North Carolina on the first map), but it acts as the local bank for communities across the nation. The question this map answers is:

"To what extent does a community rely on banks that received TARP funds?"

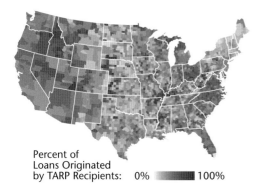

Percent of
Loans Originated
by TARP Recipients: 0% ▬▬▬ 100%

Another map created by Webb maps the percent of loans per county by banks that received TARP funds. The polarization of the first two maps disappears. Big banks created mortgages in communities they had no established relationship with while sheltering themselves from the consequences of their lending practices. The question this map answers is:

"To what extent can a bank establish itself on Main Street without even building a branch?"

Banks are not points but dynamic institutions accumulations and flows of capital. Before you try to marry it to a graphic mark, consider the concept carefully.

Visual Variables

Mappable data vary immensely. One approach to symbolizing your data, the visual variables, guides map symbolization by considering the characteristics of your data. Are your data at points, along lines, or in areas? Are your data qualitative or quantitative? Wedded to a careful consideration of the concepts behind your data, the visual variables serve to guide basic map symbol design.

Points, Lines, or Areas?

Most mappable data are at points (zero dimensions), lines (one dimension), or in areas (two dimensions).

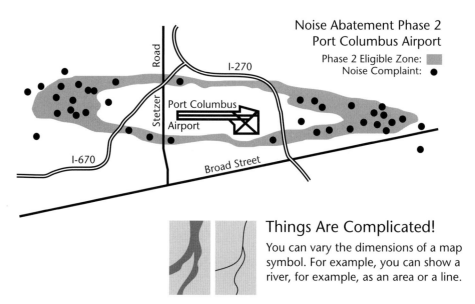

Noise Abatement Phase 2
Port Columbus Airport
Phase 2 Eligible Zone:
Noise Complaint:

Things Are Complicated!

You can vary the dimensions of a map symbol. For example, you can show a river, for example, as an area or a line.

Qualitative or Quantitative?

Consider next whether your data vary in either quality (differences in kind) or quantity (differences in amount). Some data are not easily qualitative or quantitative.

Qualitative Data

> House location
> Border or boundary
> Land vs. water
> Religious denominations
> Animal species
> Plant types
> Sexual orientation
> Political affiliation

Quantitative Data

Ordinal: distinctions of order with no measurable difference between the ordered data: high-, medium-, and low-risk zones

Interval: distinctions of order with measurable differences among the ordered data but no absolute zero: temperature F°

Ratio: distinctions of order with measurable differences between the ordered data and an absolute zero: murder rate per country

Matching Data to Visual Variables

Particular visual variables may suggest important characteristics of your data. If your data are qualitative, choose a visual variable that suggests qualitative differences, such as shape or color hue. If your data are quantitative, choose a visual variable that suggests quantitative differences, such as size or color value. Some visual variables can be manipulated to suggest either qualitative or quantitative differences, such as texture.

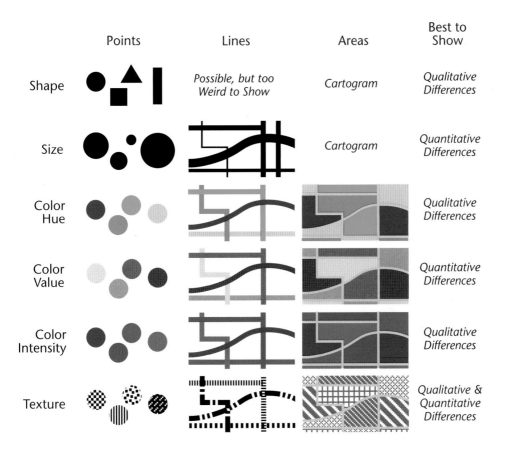

	Points	Lines	Areas	Best to Show
Shape		Possible, but too Weird to Show	Cartogram	Qualitative Differences
Size			Cartogram	Quantitative Differences
Color Hue				Qualitative Differences
Color Value				Quantitative Differences
Color Intensity				Qualitative Differences
Texture				Qualitative & Quantitative Differences

Shape

Map symbols with different shapes imply differences in quality. A square is not more or less than a circle, but is different in kind. Map symbol shapes can be pictorial or abstract.

☹ use of shape

Active Hate Groups, 2010

★ 41 – 84
♦ 21 – 40
● 11 – 20
■ 5 – 10
▲ 0 – 5

Shape is a poor choice for showing *quantitative* data. Using shape makes it hard to see the patterns on the map, as the symbols do not suggest the order (low to high) in the data.

☺ use of shape

🔺 KKK
卐 Neo-Nazi
✊ Black Separatist
▨ Neo-Confederate
✝ Christian Identity
☠ Racist Skinhead

Shape is a good choice for showing *qualitative* data. Different shapes suggest the qualitatively different groups.

Dominant Hate Group, 2010

Size

Map symbols with different sizes imply differences in quantity. A larger square implies greater quantity than a smaller square.

☺ use of size

Active Hate Groups, 2010

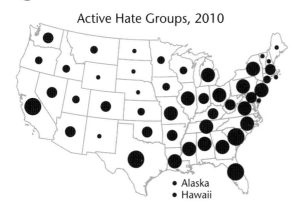

● 41 – 84
● 21 – 40
● 11 – 20
• 5 – 10
· 0 – 5

Size is a good choice for showing *quantitative* data. The use of one symbol varying in size parallels the order in the data.

☹ use of size

- · KKK
- • Neo-Nazi
- ● Black Separatist
- ● Neo-Confederate
- ● Christian Identity
- ● Racist Skinhead

Size is a poor choice for showing *qualitative* data. Different sizes suggest order in the data rather than the qualitatively different groups.

Dominant Hate Group, 2010

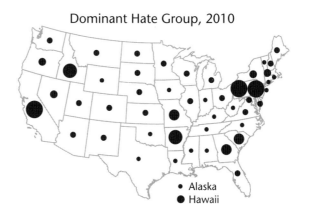

Color Hue

Color hue refers to different colors such as red and green. Symbols with different hues readily imply differences in quality. Red is not more or less than green, but is different in kind.

🙁 use of color hue

Total Iraq & Afghanistan War Casualties
Casualties per million population, as of Feb. 2010

■	80 – 230
■	42 – 79
■	26 – 41
■	13 – 25
■	16 – 12

Hue is a poor choice for showing *quantitative* data. Using hue makes it difficult to see the patterns on the map, as the colors do not suggest the order (low to high) in the data.

🙂 use of color hue

■ McCain Win
■ Obama Win

Color hue is a good choice for showing *qualitative* data. Qualitatively different hues parallel the qualitatively different data.

Presidential Election, 2008

Color Value

Color value refers to different shades of one hue, such as dark and light red. Map symbols with different values readily imply differences in quantity. Dark red is more than light red.

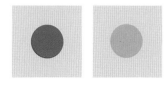

 use of color value

Total Iraq & Afghanistan War Casualties
Casualties per million population, as of Feb. 2010

■	80 – 230
■	42 – 79
■	26 – 41
■	13 – 25
■	16 – 12

Value is a good choice for showing *quantitative* data. The use of one hue varying in value parallels the order in the data.

☹ use of color value

■	McCain Win
■	Obama Win

Color value is a poor choice for showing *qualitative* data. Values suggest an ordered difference, which is not appropriate for these data.

Presidential Election, 2008

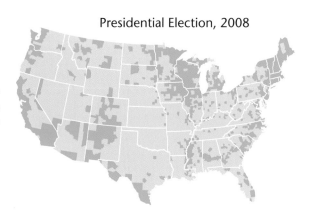

Color Intensity

Color intensity (or saturation) is a subtle visual variable that is best used to show subtle data variations, such as binary (yes or no) data that are not really qualitative or quantitative.

☹ use of color intensity

Iraq & Afghanistan War Casualties
Casualties per million population, as of Feb. 2010

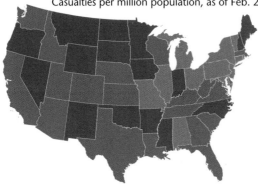

■	80 – 230
■	42 – 79
■	26 – 41
■	13 – 25
■	16 – 12

Intensity is a poor choice for showing *quantitative* data. Intensity may suggest order, but due to the lack of variation in value the sense of order is weak.

☺ use of color intensity

■ No African Americans
■ One or More African Americans

Intensity is a good choice for showing *binary* (yes/no) data. Intensity, like binary data, is neither qualitative nor quantitative.

African-American Absence, 2000

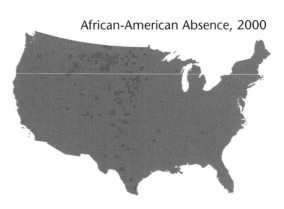

Texture

Texture (pattern) can imply both qualitative (brick vs. cloth) and quantitative (coarse vs. fine) differences. Select textures so that they suggest the qualitative or quantitative character of your data.

☹ use of texture

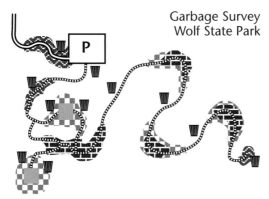

Garbage Survey
Wolf State Park

▨ Cigarette Butts
▩ Paper Debris
▦ Glass & Cans

Textures can be visually noisy and imply ordered differences. Be careful with textures that look like something: glass and cans are shown with a brick pattern, which does not make sense.

☺ use of texture

▨ Cigarette Butts
⁄ Paper Debris
▦ Glass & Cans

Texture can be good for showing qualitative data. Select textures that are not visually noisy and that suggest the qualitative differences in the data.

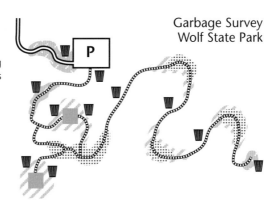

Garbage Survey
Wolf State Park

The wind barb symbols are multivariate, showing wind speed, direction, and cloud-cover. The symbol orientation, a subset of **shape**, shows wind direction (qualitative).

The **size** and number of the wind barb tails shows wind speed (quantitative).

The **value** of the wind barb circle, empty, half full, or full, suggests amount of cloud-cover (quantitative).

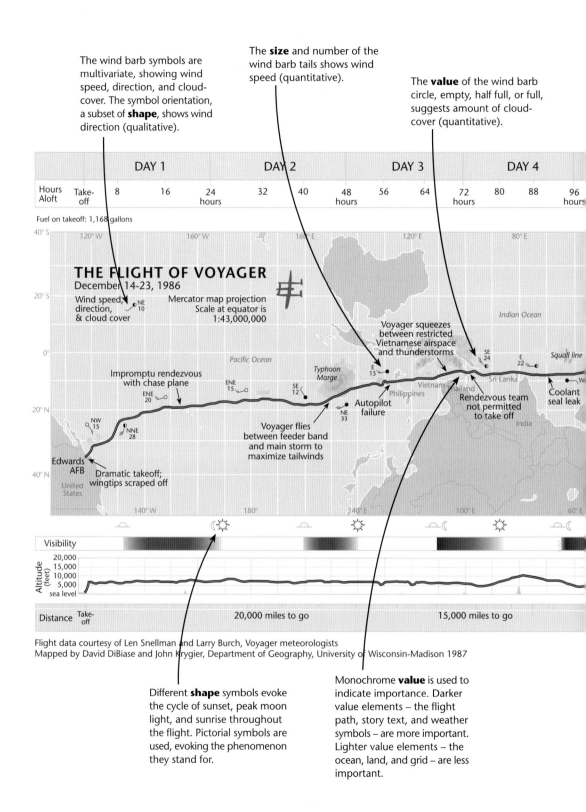

		DAY 1			DAY 2			DAY 3			DAY 4	
Hours Aloft	Take-off	8	16	24 hours	32	40	48 hours	56	64	72 hours	80	88	96 hours

Fuel on takeoff: 1,168 gallons

THE FLIGHT OF VOYAGER
December 14-23, 1986

Wind speed, direction, & cloud cover — NE 10

Mercator map projection
Scale at equator is 1:43,000,000

Indian Ocean

Pacific Ocean

Typhoon Marge

Voyager squeezes between restricted Vietnamese airspace and thunderstorms

SE 24   E 22   Squall line

Sri Lanka

Coolant seal leak

Impromptu rendezvous with chase plane   ENE 15   SE 12   E 15   Vietnam Thailand   Rendezvous team not permitted to take off

ENE 20

Philippines

Autopilot failure

NE 33

India

NW 15

NNE 28

Voyager flies between feeder band and main storm to maximize tailwinds

Edwards AFB

Dramatic takeoff; wingtips scraped off

United States

Visibility

Altitude (feet)
20,000
15,000
10,000
5,000
sea level

Distance   Take-off   20,000 miles to go   15,000 miles to go

Flight data courtesy of Len Snellman and Larry Burch, Voyager meteorologists
Mapped by David DiBiase and John Krygier, Department of Geography, University of Wisconsin-Madison 1987

Different **shape** symbols evoke the cycle of sunset, peak moon light, and sunrise throughout the flight. Pictorial symbols are used, evoking the phenomenon they stand for.

Monochrome **value** is used to indicate importance. Darker value elements – the flight path, story text, and weather symbols – are more important. Lighter value elements – the ocean, land, and grid – are less important.

The **size** of the type suggests importance (quantitative). Larger-size type labels the more important phenomena, smaller type the less important.

The **shape, size,** and **value** of the flight path – wide with a white core – suggests the symbol is very important. The distinctive symbol shape, size, and value also tie the symbol on the main map to the symbol on the altitude map.

Value is used in the top and bottom data bar to provide overall balance and stability for the map. The gray tones provide a solid base and cap to the overall map.

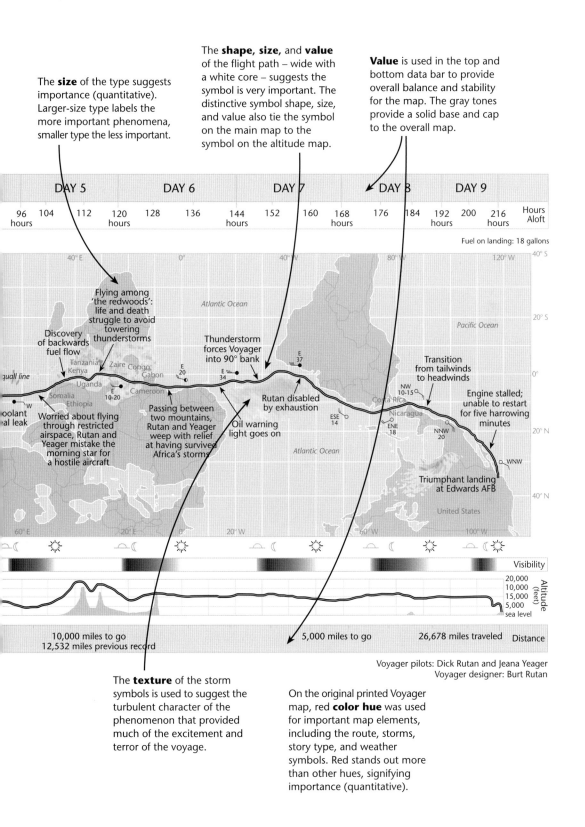

DAY 5	DAY 6	DAY 7	DAY 8	DAY 9	

| 96 hours | 104 | 112 | 120 hours | 128 | 136 | 144 hours | 152 | 160 | 168 hours | 176 | 184 | 192 hours | 200 | 216 hours | Hours Aloft |

Fuel on landing: 18 gallons

40° E   0°   40° W   80° W   120° W   40° S

Atlantic Ocean

20° S

Pacific Ocean

Flying among
'the redwoods':
life and death
struggle to avoid
towering
thunderstorms

Discovery
of backwards
fuel flow

Thunderstorm
forces Voyager
into 90° bank

Transition
from tailwinds
to headwinds

Tanzania Zaire Congo
Kenya   Gabon   E
squall line   20   E   34   E 37

Uganda   NW
Somalia   10-15
Ethiopia   E   Cameroon   Rutan disabled   Costa Rica
W   10-20   by exhaustion   Engine stalled;
coolant   Nicaragua unable to restart
al leak   Worried about flying   Passing between   Oil warning   ESE   ENE   for five harrowing
          through restricted   two mountains,   light goes on   14   18   minutes
          airspace, Rutan and   Rutan and Yeager   NNW
          Yeager mistake the   weep with relief   20   20° N
          morning star for   at having survived
          a hostile aircraft   Africa's storms   Atlantic Ocean   WNW

Triumphant landing
at Edwards AFB

United States   40° N

60° E   20° E   0°   20° W   60° W   100° W

Visibility

20,000
10,000
15,000 Altitude (feet)
5,000
sea level

10,000 miles to go   5,000 miles to go   26,678 miles traveled   Distance
12,532 miles previous record

Voyager pilots: Dick Rutan and Jeana Yeager
Voyager designer: Burt Rutan

The **texture** of the storm symbols is used to suggest the turbulent character of the phenomenon that provided much of the excitement and terror of the voyage.

On the original printed Voyager map, red **color hue** was used for important map elements, including the route, storms, story type, and weather symbols. Red stands out more than other hues, signifying importance (quantitative).

Symbolizing Aggregate Data

Symbolizing data that have been grouped into geographic areas (countries, states, etc.) is complicated: the same data can be mapped as a choropleth, graduated symbol, dot, or surface map. Choosing among these methods requires understanding what you are mapping and your goals for the map.

Incidence of AIDS in Pennsylvania, 1985

One of the most common types of mappable data is associated with geographic areas: one data value is associated with each geographic area. In the set of maps that follow, each county in Pennsylvania has one value, representing total AIDS cases; the same data are mapped with four methods. AIDS is a contagious disease unevenly spread through geographic space. Available AIDS data usually consist of a single value for a county, state, or country.

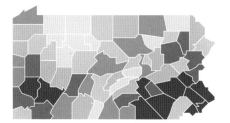

Choropleth

Darker means more (lighter, fewer) cases.

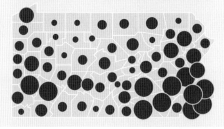

Graduated Symbol

Larger circles mean more (smaller, fewer) cases.

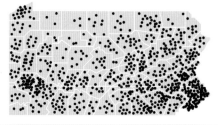

Dot

Density of dots reveals density of cases.

Surface

Darker means more (lighter, fewer) cases.

A **choropleth map** varies the shading of each area in tandem with the data value associated with it. This map suggests that the incidence of AIDS is uniform throughout each county, with potentially abrupt changes at county boundaries.

This map is useful for an AIDS educator working at the county level – suggesting that AIDS is everywhere in the county and everyone must take precautions – or for county representatives seeking aid for AIDS care from a legislature.

A **graduated symbol map** varies the size of a symbol centered on each area in tandem with the data value associated with it. This map suggests a single data value for each county. This map does not suggest a uniform distribution of AIDS cases within each county.

This map is useful for the official yearly health report for the State of Pennsylvania. The map indicates that there are AIDS cases, but the individual symbols subtly suggest AIDS is contained and under control in most of the counties.

A **dot map** varies the number of dots in each area in tandem with the data value associated with it. On this map, one dot equals 30 AIDS cases. The location of dots does not show the specific location of AIDS cases; rather, the density of dots in each area represents more or fewer AIDS cases in the area.

This map looks like it is showing specific cases of AIDS and that AIDS cases are spread throughout each county in an even manner. As neither is actually true, this map is a bit dangerous and is best used as an illustration of the dot map technique in cartographic textbooks.

The **surface map** creates an abstract surface from the single data value for each area. Imagine a pin stuck in each county, where the height of the pin varies with the number of AIDS cases. Now imagine a surface created by a sheet laid over these pins. Ta-da!

This map suggests that AIDS is everywhere in the state and highly contagious. This map is, then, useful for a statewide AIDS awareness campaign intended to scare randy teens.

Choropleth Maps

The choropleth map is one of the most common mapping techniques for data grouped into areas – counties, provinces, states, countries, etc. Choropleth maps vary the shading of each area along with the data value associated with the area.

Appropriate Data: Derived Data (Density, Rates), Sometimes Totals

Mapping total numbers with a choropleth map is usually not recommended, especially when the areas on the map vary in size. A large area may have more people simply because it covers a larger area.

If you map totals (bold, below) and classify the data, an area with 100 people will likely be in a different class than the area with 500 people. The visual difference between the areas on the map is the result of the unequal size of the areas.

Mapping derived data, like population per square mile, takes into account the varying size of areas on the map.

If you map densities (bold, below) and classify the data, both areas have 10 people per square mile and will be in the same class.

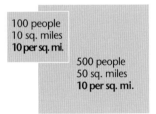

Mapping totals with a choropleth map can be OK! A marketing company maps the total number of Polish-speaking U.S. citizens by county in the U.S., to assist in a plan to market Polish greeting cards. In this case, what is most important is where the most Polish-speaking folks are. But do consider a graduated symbol map for totals. It may better serve your needs.

The goals for your map drive your choices! There is no absolute "best" kind of data for a choropleth map independent of your goals for the map. Be aware of the problem of mapping totals with a choropleth map, but if your goals require totals, just do it. Then create a graduated symbol map of the same data, and compare the maps. Please use your brain when making maps.

Choropleth Map Design: Value, Legend, and Boundaries

☹ value and legend

Wisconsin Farm Density

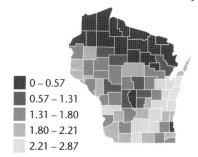

- ■ 0 – 0.57
- ■ 0.57 – 1.31
- ■ 1.31 – 1.80
- ■ 1.80 – 2.21
- ■ 2.21 – 2.87

Dark means less is unconventional
Smaller values at legend top are unconventional
A continuous legend is used; there is no gap
  between class breaks, falsely suggesting
  that values span the full range of the class

☹ value and boundaries

Wisconsin Farm Density

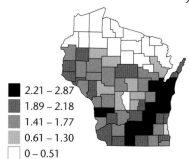

- ■ 2.21 – 2.87
- ■ 1.89 – 2.18
- ■ 1.41 – 1.77
- ■ 0.61 – 1.30
- □ 0 – 0.51

Black for top class blends with boundaries
White for bottom class suggests no data
Boundaries jump out too much

☺ value and legend and boundaries

Wisconsin Farm Density
Farms per
sq. mile
2000

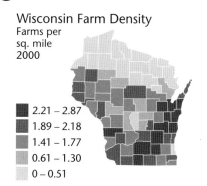

- ■ 2.21 – 2.87
- ■ 1.89 – 2.18
- ■ 1.41 – 1.77
- ■ 0.61 – 1.30
- ■ 0 – 0.51

Dark means more to most people
A noncontinuous legend provides an indication
  of the actual data variation in each class by
  showing the high and low value
Larger values at the legend top (high = more)
A title that explains the numbers in the legend
Overall smoother feel without black and white
  class shading, while removing the problems
  they cause
Boundaries less dominant but distinct

Graduated Symbol Maps

The graduated symbol map varies the size of a single symbol, placed within each geographic area, in tandem with the data value associated with the area.

Appropriate Data: Totals, Sometimes Derived Data (Densities, Rates)

Mapping derived data with graduated symbols is usually not recommended. Graduated symbols readily imply magnitude rather than density or rates.

If you map density (bold, below), the symbols imply no magnitude difference in population between the two areas.

Use totals for graduated symbol maps. Mapping totals with graduated symbols suggests the magnitudes inherent in the data.

If you map totals (bold, below) the map implies a difference in total population between the two areas.

100 people
10 sq. miles
10 per sq. mi.

500 people
50 sq. miles
10 per sq. mi.

100 people
10 sq. miles
10 per sq. mi.

500 people
50 sq. miles
10 per sq. mi.

Mapping derived data with a graduated symbol map can be OK! A map of global coffee consumption might, for example, use graduated coffee cups to show the percentage of total coffee consumption in each country then relate the map to a pie chart showing the same data. Coffee and pie, delicious.

Consider a choropleth map for these data; it may serve your needs better. But the graduated coffee cups may be too sweet to pass up.

The goals for your map drive map choices! There is no absolute "best" kind of data for a graduated symbol map – independent of your goals. Be aware of the problem of mapping derived data with a graduated symbol map, but if your goals require it, it's OK to do it.

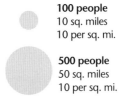

Graduated Symbol Map: Classification

Classified Legend

⬤ 8,001 – 10,000 persons

⬤ 5,001 – 8,000 persons

● 1,001 – 5,000 persons

• Less than 1,000 persons

Classified graduated symbol maps use standard classification schemes. Assign one symbol size for each class.

Less data detail
Easier to match particular symbol on map to legend
Easier to see distinct classes in data

Unclassified Legend

⬤ 9,000 persons

● 6,500 persons

● 2,500 persons

• 500 persons

Unclassified graduated symbol maps scale each symbol to each value. Legend should include *representative* symbol sizes.

More data detail
Harder to match particular symbol on map to legend
Harder to see distinct classes in data

Graduated Symbol Map: Symbol Design

Squares, Triangles

Less compact symbol
Edgy visual impression good for edgy phenomena

Circles

More compact symbol
Smooth visual impression good for mellow phenomena

Volumetric Shapes

Visually attractive
Suggest volumetric phenomena, avoid these stinkers

Pictographic Shapes

Visually attractive
Easy to understand
Potentially cute and distracting

Cartograms

The cartogram is a variant of the graduated symbol map. Cartograms vary the size of geographic areas (rather than symbols) based on the single data value associated with the area. Cartograms, while difficult to create, are visually striking.

Appropriate Data: Totals and Derived Data (Densities, Rates)

Derived Data: U.S. Suicide Rate

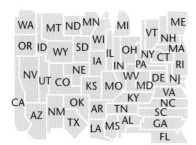

Cartograms of data with minimal variation from area to area can be interesting. Suicide rates don't vary much around the U.S., and thus all the states are about the same size.

Cartograms are not effective when the map reader is not familiar with the geographic areas being varied.

Total Data: 2008 U.S. Election

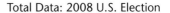
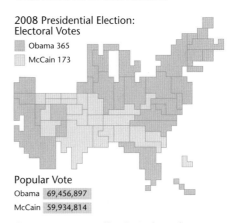

2008 Presidential Election: Electoral Votes

Obama 365

McCain 173

Popular Vote

Obama 69,456,897

McCain 59,934,814

Cartograms are effective when data variation from area to area is significant. A cartogram scaled to electoral votes in each state in the U.S. removes the confusion introduced by state size.

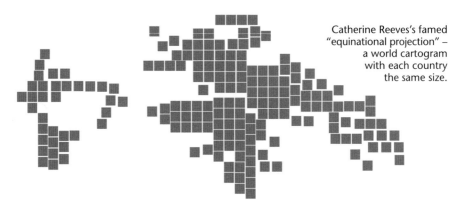

Catherine Reeves's famed "equinational projection" – a world cartogram with each country the same size.

Cartogram: Form

Apportionment Map
William Bailey, 1911

Apportionment means "allotment in proper shares." Each area's size is based on population, not geographic area. Bailey's cartogram is contiguous, roughly retaining adjacency. Compare to noncontiguous cartograms, such as the "equinational projection" (preceding page).

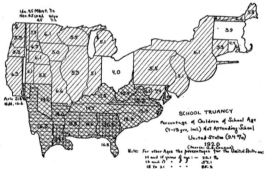

Population Projection
Karl Karsten, 1923

"The corrected areas of the States serve to give an excellent background or evaluation of the importance of the statistics plotted upon the map." Karsten suggested this map be sold, as a blank map, for compiling a second data variable, such as truancy (left), thus allowing the creation of a bivariate cartogram.

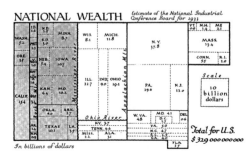

Rectangular Statistical Cartogram
Erwin Raisz, 1934

"It should be emphasized that the statistical cartogram is not a map. The cartogram is purely a geometrical design to visualize certain statistical facts and to work out certain problems of distribution." Raisz's cartogram was similar in form to early European cartograms.

Dot Maps

The dot map varies the number of dots in each geographic area, based on the single data value associated with the area. Dots on a dot map do not represent the specific location of a single instance of some phenomena; rather, the density of dots in a geographic area represents the density of phenomena in that area.

Appropriate Data: Totals, Not Derived Data (Densities, Rates)

Using a dot map for derived data is not recommended. "One dot equals 50 people per square mile" is too weird to think about. Use totals instead. Each dot equals a number of phenomena.

Mapping software randomly locates dots in areas on the map, which is misleading. Random concentrations suggest patterns that do not exist. The larger the geographic area, the more likely randomly generated patterns will appear.

Random dot placement on the map below suggests concentrations of Rastafarians in rural northern areas of these counties, which is incorrect. Rastafarians mostly live in the urban cultural centers of Brock, Athens, and Devin.

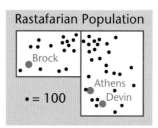

Make your dot map using smaller geographic areas (below). Create a dot map based on data for smaller areas, then remove the smaller area boundaries.

Dot placement should be guided by additional knowledge or filters in mapping software. Thus more of the dots are placed around urban areas than in rural areas.

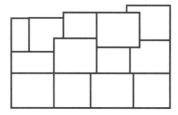

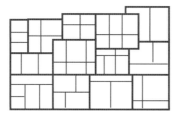

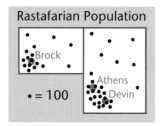

194

Dot Map: Classification, Dot Value, and Size

☹ dot value (too dense)

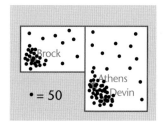

• = 50

☺ dot value (value changed)

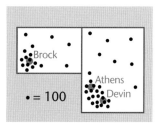

• = 100

☹ dot value (too sparse)

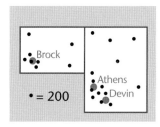

• = 200

☺ dot value (size change)

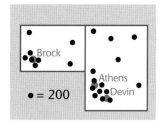

• = 200

Adjust dot value and/or size so that dots begin
  to coalesce in densest areas
Use a round number for dot value (100, not 142)

Anti-Dotite Diatribe: Old school dot maps made by folks familiar with the geographic context of their data placed dots where the phenomena were likely to be found. Unless you are very familiar with your data's context and can manipulate dot placement, you should avoid using dot maps.

Anti-Dotite Diatribe (continued): Most normal people will think the mapped dots refer to a single instance of the phenomena at their actual locations, which is not the case. Unless your map's viewers are very familiar with the dot map method, you should avoid using dot maps.

Surface Maps

Surface maps create an abstract 3D surface (isoplethic) based on one data value associated with each area. Surface maps can also be created from data at points (isometric). Isoplethic maps, the focus here, suggest continuous phenomena.

Appropriate Data: Derived Data (Densities, Rates)

Using a surface (isopleth) map for totals is often not recommended, particularly when the areas on the map vary in size. A large area may have more people simply because it covers a larger area.

Use derived data such as densities for surface (isopleth) maps. Mapping people per square mile takes into account the varying size of areas on the map so the map user can see real differences in the distributions.

If you map totals (below), an area with 100 people will likely be at a different level than an area with 500 people. The visual difference between the areas is the result of the unequal size of the areas.

If you map densities (below), both areas have 10 people per square mile and will be at the same level. The visual difference (or lack of difference) between the areas is the result of the data.

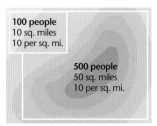

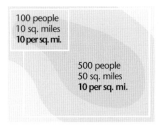

Mapping totals with a surface map can be OK! For example: a hunter wants to shoot as many chupacabras as possible, due to a $20-per-head bounty. A surface map of estimated chupacabras per county will guide him to where more of these goat-suckers might be found. Do consider a graduated symbol map for totals; it may better serve your needs.

Your goals for your map drive map choices! There is no absolute "best" kind of data for a surface map – independent of your goals for your map. Be aware of the problem of mapping totals with a surface map, but if your goals require totals, just do it. Then create a graduated symbol map of the same data, and compare the maps.

Surface Map: Shading

Contour lines show constant values slicing through the abstract 3D surface. Contour lines may produce a busy map with little room for other data. They may also be misinterpreted as a tangible linear feature by naive map readers.

Filled contours are less busy than contour lines and allow other data to be super-imposed. They also are more suggestive of a surface. Follow the convention of dark equals more.

 design

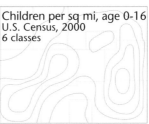

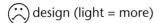 design (light = more)

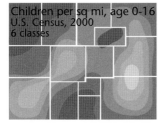

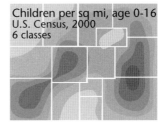 design (dark = more)

Surface Map: Legend

 legend

Contour Interval =
50 people per sq. mile

legend

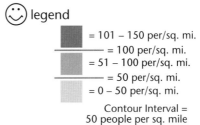

= 101 – 150 per/sq. mi.
= 100 per/sq. mi.
= 51 – 100 per/sq. mi.
= 50 per/sq. mi.
= 0 – 50 per/sq. mi.

Contour Interval =
50 people per sq. mile

What do I care for the colored pins on a General's map? It' s not a fair bargain – this exchange of my life for a small part of a colored pin.

Irwin Shaw, *Bury the Dead* (1936)

But to look at the stars always makes me dream, as simply as I dream over the black dots of a map representing towns and villages. Why, I ask myself, should the shining dots of the sky not be as accessible as the black dots on the map of France? If we take the train to get to Tarascon or Rouen, we take death to reach a star. One thing undoubtedly true in this reasoning is this: that while we are alive we cannot get to a star, any more than when we are dead we can take the train.

Vincent van Gogh (1888)

And then I went to bed, and went to sleep, and slept soundly, and the next morning I sent for the chief engineer of the War Department (our map-maker), and I told him to put the Philippines on the map of the United States (pointing to a large map on the wall of his office), and there they are, and there they will stay while I am President!

President William McKinley (1899)

▲   Group of Boulders

This symbol represents a group of boulders. You read that right. This TRIANGLE represents a group of BOULDERS. BOULDERS. A GROUP OF THEM. I don't like this symbol. Why not make something more boulder-ish? A TRIANGLE? REALLY?!?!

A triangle.

A fucking triangle.

Student answer, map symbol critique on exam, Geography 222, Ohio Wesleyan University (2010)

More...

Symbolization occupies large sections of any of the previously cited texts. A broader semiological approach to map and graphic symbols can be found in the work of Jacques Bertin, *Graphics and Graphic Information Processing* (Walter de Gruyter, 1981) and *Semiology of Graphics* (ESRI Press, 2010). Eduard Imhof's *Cartographic Relief Presentation* (ESRI Press, 2007) is the classic text on terrain symbolization.

For early examples of many of the map symbolization methods reviewed in this chapter, see Arthur Robinson's *Early Thematic Mapping in the History of Cartography* (University of Chicago Press, 1982). The magisterial volumes in the *History of Cartography* series, published by the University of Chicago Press, provide examples of a diversity of maps from different cultures throughout time.

A neat little tome on isotype is Marie Neurath and Robin Kinross, *The Transformer: Principles of Making Isotype Charts* (Hyphen Press, 2009).

Two fun collections of maps by artists – revealing how they have appropriated map symbolization for their own purposes – are Katharine Harmon's *You Are Here: Personal Geographies and Other Maps of the Imagination* (Princeton Architectural Press, 2004) and *The Map as Art,* cited at the end of Chapter 2.

Sources: The isotype map is from *Report of the Mississippi Valley Committee of the Public Works Administration* (U.S. Government Printing Office, 1934, p. 8). The 17th-century Russian map is from *Rossiiskii Gosudarstvennyi Arkhiv Drevnikh Aktov,* Moscow (n.d.). The prioria copaifera map is from *Smithsonian Tropical Research Institute.* The 18th-century Russian tree map is from Liudmila Shaposhnikova, *Izobrazhenie lesa na Kartakh* (Moskva, Izd-vo Akademii nauk SSSR, 1957). The West Sussex tree is from Kim Leslie, *A Sense of Place: West Sussex Parish Maps* (West Sussex County Council, 2006). Richard Florida's TARP map is redrawn from the original posted on *The Atlantic* website (August 20, 2009), itself based on data from propublica.org. Kevin Webb's maps and data, from which we redrew our two county-level TARP maps, are at subsidyscope.com. Hate group data are from the *Southern Poverty Law Center* (www.splcenter.org). Iraq and Afghanistan war casualties are from U.S. Department of Defense Personnel and Procurement Statistics. Maps of Pennsylvania AIDS data are redrawn from maps in Alan MacEachren, *Some Truth with Maps* (Association of American Geographers, 1994). The suicide rate cartogram is redrawn from Dent's *Cartography: Thematic Map Design* (2008). The "Equinational" cartogram, designed by Catherine Reeves, is redrawn from *Globehead! The Journal of Extreme Geography* (1994). Bailey's "apportionment map" was published on April 6, 1911, in *The Independent.* Karsten's "population projection" was published as a mimeographed sheet and also reproduced in his *Charts and Graphs* (Prentice Hall, 1925). Erwin Raisz's rectangular statistical cartogram was published in "The Rectangular Statistical Cartogram," *Geographical Review,* v. 24, no. 2 (April 1934), p. 294.

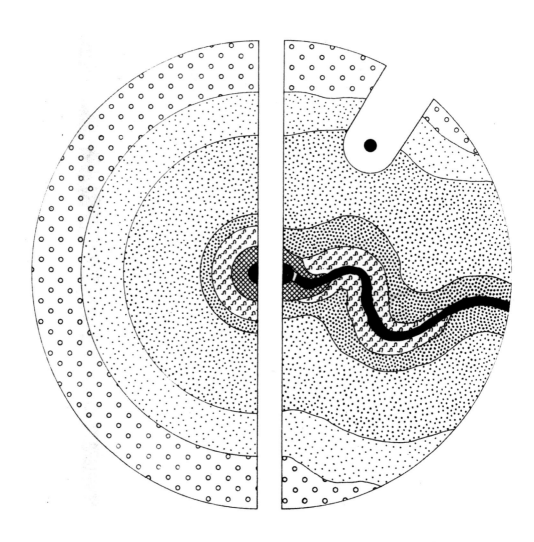

St.

SACRE COEUR

PARC DE
MONCEAU

GARE ST.
LAZARE

MONTMARTRE

CHAMPS
ELYSEES

OPERA

MADELEINE

GRANDS
BOULEVARD

DEFENSE

ETOILE

CONCORDE

TUILERIES

LOUVRE

PALAIS
ROYALE

GRAND PALAIS

QUAIS ET BERGES

SEINE

CHAILLOT

TOUR EIFFEL

CHAMBRE
DES DEPUTES

ST. GERMAIN

NOTRE
DAME

BOIS DE
BOULOGNE

CHAMPS DE MARS

INVALIDES

LUXEMBOURG

ST. MICHEL

ECOLE MILITAIRE

JARDIN DES
PLANTES

PANTHEON

MONTPARNASSE

QUARTIER LATIN

GARE
D'AUSTERLITZ

ITALIE

PARC DE
MONTSOURIS

BUTTES CHAUMONT

GARE DU
NORD

GARE DE
L'EST

REPUBLIQUE   PERE LACHAISE

LES-HALLES

PLACE DES VOSGES

MARAIS

BASTILLE

ST. LOUIS

NATION

GARE
DE LYON

BOIS DE
VINCENNES

How can the words mean
more than they say?

SACRE COEUR
23.4%

PARC DE
MONCEAU
17%

GARE ST.
LAZARE
16.6%

MONTMARTRE
32.1%

MADELEINE
17.9%

OPERA
30.7%

GRANDS
BOULEVARD
9.2%

DEFENSE
9.7%

CHAMPS
ELYSEES
40.4%

CONCORDE
45.4%

TUILERIES
33.5%

LOUVRE
45.4%

PALAIS
ROYALE
15.2%

ETOILE
61.9%

GRAND PALAIS
9.7%

QUAIS ET BERGES
22.5%

SEINE
84.3%

CHAILLOT
32.1%

TOUR EIFFEL
54.6%

CHAMBRE
DES DEPUTES
11.5%

ST. GERMAIN
31.2%

NOTRE
DAME
55.5%

BOIS DE
BOULOGNE
49.1%

CHAMPS DE MARS
17.9%

INVALIDES
29.8%

LUXEMBOURG
38.5%

ST. MICHEL
30.1%

JARDIN DES
PLANTES
16.1%

ECOLE MILITAIRE
11.5%

Seine

PANTHEON
20.7%

MONTPARNASSE
35.3%

QUARTIER LATIN
20.7%

GARE
D'AUSTERLITZ
13.8%

ITALIE
12.4%

PARC DE
MONTSOURIS
16.6%

BUTTES CHAUMONT
24.4%

GARE DU
NORD
14.7%

GARE DE
L'EST
15.6%

REPUBLIQUE   PERE LACHAISE
14.3%   12.9%

LES-HALLES
10.1%
        PLACE DES VOSGES
        18.4%

MARAIS
26.2%   BASTILLE
                   22.1%

ST. LOUIS
31.7%

                NATION
                12%

GARE
DE LYON
18.4%

Seine

        BOIS DE
        VINCENNES
        38.1%

CHAPTER

10 Words on Maps

Words on maps mean what they say and also mean what they show. A word alone means something, but on this map its size also means something. Stanley Milgram was a psychologist interested, among other things, in the mental images people made of their environment. He asked over 200 Parisians to draw maps of Paris. This map shows the 50 most cited elements. The name of each locale is shown in a size proportional to the number of subjects who included it in their hand-drawn maps of Paris. In this case, type size makes the point: the real magic is in the names.

What Words Mean

It's names that make a map a map. Without names maps fade into pictures or photographs. They lose their great power, which is to christen and claim. Named and renamed, no part of the world has escaped the map's heavy hand. Every inflection of the land, every body of water, has been categorized and christened. Every path, byway, roadway, and street, every hamlet, village, town, and city has been tagged.

Martin Waldseemüller's 1507 map named the continents west of the Atlantic "America." Americans have been Americans ever since.

A map of autumn leaf color in the Boylan Heights neighborhood in Raleigh, NC, made entirely of color names. Even by themselves, names can do a lot of work.

RED
RED RED
RED RED green
RED green OCHRE purple
buff green orange
olive olive mottled green
green OCHRE purple green purple green
olive orange yellow yellow
olive REDgreen yellow purple
olive yellow greenish yellow green mottled yellow
olive GOLD green GOLD REDgreen GOLD orange
mauve green greenish yellow green olive green
buff green purple GOLD mauve purple lemon
olive mottled green mottled olive RED ruby purple green
greenish yellow olive yellow marigold GOLD
olive green green OCHRE purple greenish yellow orange purple
olive purple buff green RED mauve purple tangerine mottled
GOLD green dappled GOLD olive yellow olive green green
purple green mottled purple yellow clotted green
greenish yellow
bruise RED greenish yellow mauve green green purple greenish yellow
RED green yellow GOLD mottled green mauve marigold purple
orange green chartreuse purple green GOLD greenish yellow RED dappled RED green
mauve mauve marigold yellow greenish yellow mottled green green
green clotted green orange greenish yellow dandelion marigold green
green marigold purple mauve green greenish yellow OCHRE clotted chartreuse
tangerine green OCHRE RED purple greenish yellow OCHRE chartreuse mauve greenish yellow green
yellow OCHRE REDpurple dappled mauve green olive
chartreuse clotted olive green clotted yellow GOLD ruby OCHRE
green orange mottled RED green purple green RED
olive purple purple green yellow RED clotted purple
yellow mottled greenish yellow OCHRE GOLD dappled yellow mottled clotted
GOLD green clotted buff green purple mauve
yellow green GOLD green OCHRE OCHRE
greenish yellow purple yellow greenish yellow clotted mauve bruise
OCHRE green mauve green dappled mottled
chartreuse green greenish yellow olive
purple GOLD mauve chartreuse olive
marigold yellow
yellow green purple
greenish yellow mottled olive
dappled green green

206

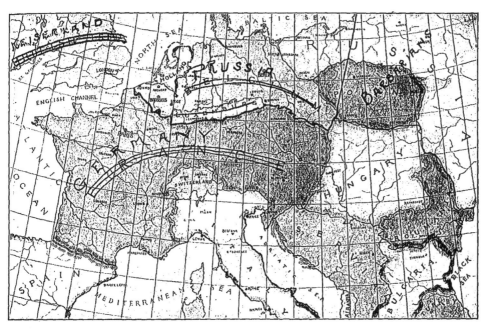

Assigning names can be contentious. In 1914 a map of Europe was published in *Life* Magazine. "An offended reader" of the magazine "corrected" the map and returned it to *Life,* where it was republished.

When the public owner of London's famous tube map demanded all spoofs of the tube map be removed from an artist's website, threatening legal action, another map began to circulate on the web.

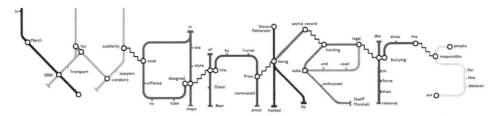

What Words Look Like: Type Anatomy

Map type is as diverse as any other symbol. Variations in type anatomy shape its look: its overall form (typeface, font), the way the strokes on letters finish (serifs), the size of the main body of lowercase letters (x-height), the parts that stick up above and below the main body (ascenders, descenders), and the size of the type.

Typeface | Font is a set of letters and numbers with a unique design. A font is a subset of a typeface, including all letters and numbers of a specific size. Font is often used to mean typeface. This particular typeface is Times Roman:

Making Maps

x-height is the height of the most compact letters in a typeface, such as an a, o, or e. Type with a greater x-height is typically easier to read.

Point size 48 point type here, where 72 points equals one inch. Type size is determined by the height of the original lead foundry block, and is not the same as the height of the letter; thus 8 point Times Roman is smaller than 8 point Helvetica.

Serifs are finishing strokes added to the ends of letters. Helvetica has no serifs (sans serif); Times Roman has them.

12 point Helvetica

14 point Helvetica

18 point Helvetica

24 point Helvetica

48 point

Typefaces | Fonts vary immensely and for mapping uses are typically divided into two broad classes – serif fonts, such as Times Roman, and sans serif fonts, such as Helvetica:

Making Maps

Sans serif

x-height

Ascender is the portion of certain letters that rises above the x-height, such as in the letters k or f.

Descender is the portion of certain letters that falls below the X-height, such as in the letters p or g.

Typefaces (fonts) are often **classified** according to their historical development, from early typefaces based on the style of hand lettering to later, much more abstract and geometric, styles.

Sabon
is a humanist or old style typeface, designed to emulate the look of calligraphy

Baskerville
is a transitional typeface, sharper and with more contrast than humanist typefaces

A **type family** is a collection of variations on a single typeface (font) appropriate for different uses. Type families may be small, or have dozens of variants.

Bodoni
is a modern typeface, abstract with strong thick/thin contrast

Clarendon
is an Egyptian or slab serif typeface, with heavy serifs, designed to be used in ads

Gill Sans
is a humanist sans serif typeface, calligraphic like humanist typefaces but lacking the serifs

Helvetica
is a transitional sans serif typeface, like transitional typefaces lacking the serifs

Futura
is a geometric sans serif typeface based on basic geometric forms

This is Bodoni standard

This is Bodoni standard bold

This is Bodoni standard bold condensed

This is Bodoni standard bold italic

This is Bodoni standard book

This is Bodoni standard book italic

This is Bodoni standard italic

This is Bodoni poster

This is Bodoni standard poster compressed

This is Bodoni standard poster italic

Type variations on your map should mean something.
Convey information with type style, size, weight, and form.

What Words Look Like: Type as Map Symbol

Map type can be used as a map symbol to differentiate qualities (typeface, type color hue, italics) and quantities and order (type size, weight, type color value) or both (type spacing, case). Make sure that variations in type on your map reflect variations in your data.

Typeface (Font): Qualities

Typeface has a significant visual impact on your map. Different typefaces can be used to suggest qualitative aspects of data and to shape the overall feel of your map.

Serif type, such as Sabon implies tradition, dignity, and solidity.

Sans Serif type, such as Stone Sans, implies newness, precision, and authority.

Typeface (Font) Considerations

Avoid combining more than two typefaces on a map unless they are designed to be used together (e.g., Stone Sans and Stone Serif)
Evaluate compatibility if different typefaces are combined
Avoid combining two serif or two sans serif type styles on one map
Serif type is easier to read in blocks of text
Decorative type styles are difficult to read and can look goofy

Type Size: Order

Type size variations imply quantitative, ordered differences. Larger sizes imply more importance or greater quantity, smaller sizes less importance or less quantity.

More important

Less important

Type Size Considerations

Type less than 6 points in size is hard to read on paper, 10 points on computer screen
Use a 2 point difference in small type sizes and 3 point difference in medium and larger type sizes if you want a noticeable difference
Most people have difficulty distinguishing more than five to seven categories of data symbolized by type size on a map
Adjust for your final medium. Increase type size for computer display or posters, decrease for print medium

Screen Fonts, Printer Fonts

Screen fonts are designed to work best on computer monitors.

Verdana is spacious with a large x-height. It may not look great on paper, but it will be more legible on a computer than Sabon.

The Stone family of typefaces (including Stone Sans) was designed to work well on lower-resolution laser printers.

Type Weight: Order

Type weight variations imply ordered (quantitative) differences. Bold type implies more importance or greater quantity, standard or light type less importance or less quantity.

More important

Less important

Type Weight Considerations

Bold implies significance and power, yet may look **pudgy**
Bold makes gray type more legible
Don't underline, use **bold** instead
Ordered categories using size and **bold**:

8 pt. standard
8 pt. bold
10 pt. standard
10 pt. bold
12 pt. standard
12 pt. bold

Type Form: Qualities, Order

Type form variations can suggest qualities (italics, color hue), order (color value), and both (spacing, case).

Water feature

BIG WATER FEATURE

Type Form Considerations

Italics is conventional for qualitative data differences, water and other natural features
Variations in the color hue (red, green, blue) of type imply qualitative data differences
Variations in the color value (light red, dark red) of type imply ordered data differences
Carefully evaluate light values of type for legibility on your final medium
E x t e n d e d or spaced type is OK for extended area features but don't go o v e r b o a r d
Condensed type may look squished
UPPER CASE implies greater quantity or importance but is more difficult to read than a mix of upper- and lower-case letters and seems a bit shouty

The *Geo-Smiley Terror Spree* map and this book use Stone Sans **typeface** (font). While designed for low-resolution laser printers, Stone works well in any medium. It is harder to screw up on the map than with fussier typefaces.

Type size clearly distinguishes the title from other map type, as well as the story text from state abbreviations. What's important is larger and more noticeable.

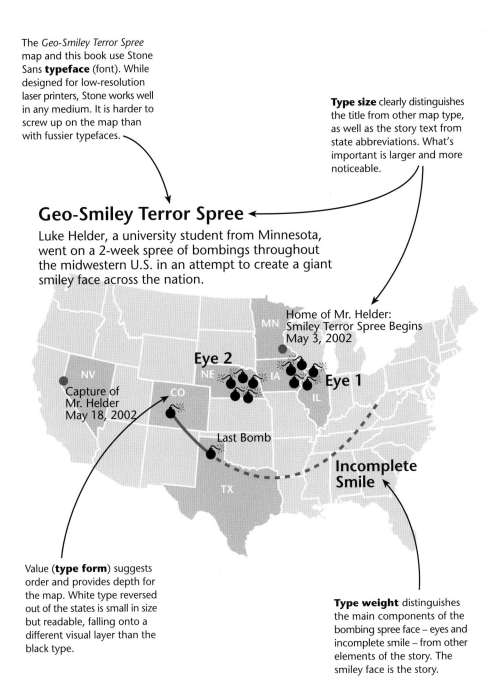

Geo-Smiley Terror Spree

Luke Helder, a university student from Minnesota, went on a 2-week spree of bombings throughout the midwestern U.S. in an attempt to create a giant smiley face across the nation.

Home of Mr. Helder: Smiley Terror Spree Begins May 3, 2002

Eye 2

Eye 1

NV

MN

NE

IA

IL

Capture of Mr. Helder May 18, 2002

CO

Last Bomb

TX

Incomplete Smile

Value (**type form**) suggests order and provides depth for the map. White type reversed out of the states is small in size but readable, falling onto a different visual layer than the black type.

Type weight distinguishes the main components of the bombing spree face – eyes and incomplete smile – from other elements of the story. The smiley face is the story.

Geo-Smiley Terror Spree

Luke Helder, a university student from Minnesota, went on a 2-week spree of bombings throughout the midwestern U.S. in an attempt to create a giant smiley face across the nation.

Sabon, an old style serifed **typeface** (font) designed to look a bit like calligraphy, is too elegant and traditional a typeface choice for this goofy, modern story.

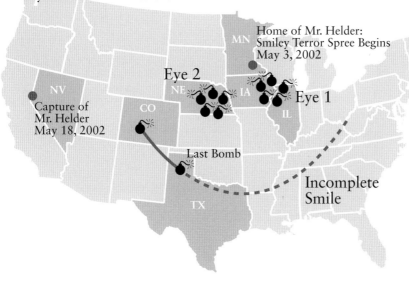

Geo-Smiley Terror Spree

Luke Helder, a university student from Minnesota, went on a 2-week spree of bombings throughout the midwestern U.S. in an attempt to create a giant smiley face across the nation.

Futura, a modern, geometric sans serif **typeface** (font), is better than Sabon but is slightly distracting: the look of the typeface is somewhat more noticeable than it could be.

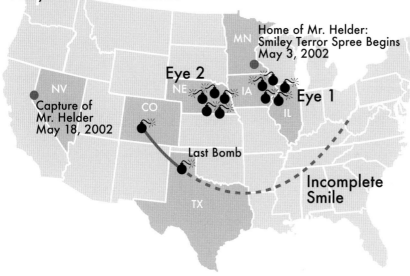

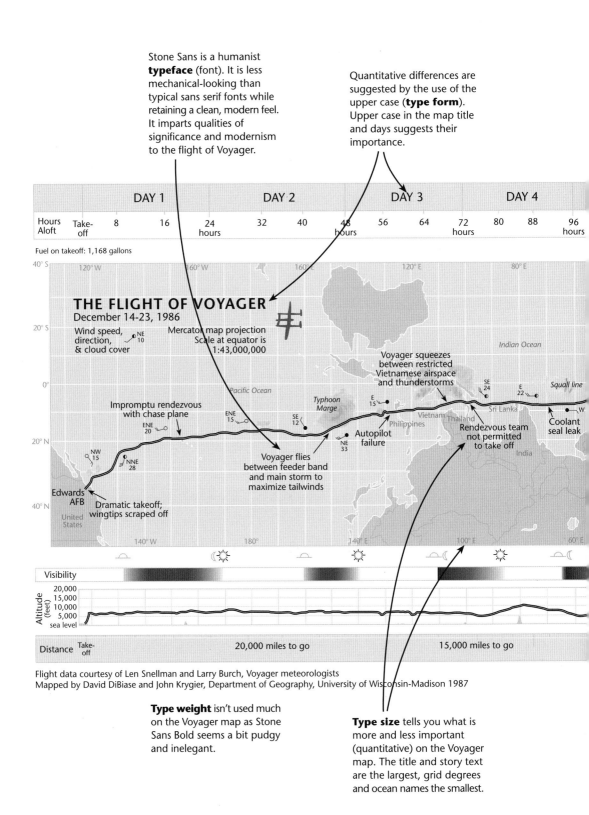

Stone Sans is a humanist **typeface** (font). It is less mechanical-looking than typical sans serif fonts while retaining a clean, modern feel. It imparts qualities of significance and modernism to the flight of Voyager.

Quantitative differences are suggested by the use of the upper case (**type form**). Upper case in the map title and days suggests their importance.

		DAY 1			DAY 2			DAY 3			DAY 4	
Hours Aloft	Take-off	8	16	24 hours	32	40	48 hours	56	64	72 hours	80	88	96 hours

Fuel on takeoff: 1,168 gallons

THE FLIGHT OF VOYAGER
December 14-23, 1986

Wind speed, direction, & cloud cover

Mercator map projection
Scale at equator is
1:43,000,000

Pacific Ocean

Indian Ocean

Typhoon Marge

Voyager squeezes between restricted Vietnamese airspace and thunderstorms

Squall line

Impromptu rendezvous with chase plane

ENE 20

ENE 15

SE 12

E 15

SE 24

E 22

W

Sri Lanka

Coolant seal leak

Vietnam *Thailand*

Philippines

Rendezvous team not permitted to take off

NE 33

Autopilot failure

Voyager flies between feeder band and main storm to maximize tailwinds

NW 15

NNE 28

India

Edwards AFB

Dramatic takeoff; wingtips scraped off

United States

Visibility

Altitude (feet): 20,000 / 15,000 / 10,000 / 5,000 / sea level

Distance: Take-off | 20,000 miles to go | 15,000 miles to go

Flight data courtesy of Len Snellman and Larry Burch, Voyager meteorologists
Mapped by David DiBiase and John Krygier, Department of Geography, University of Wisconsin-Madison 1987

Type weight isn't used much on the Voyager map as Stone Sans Bold seems a bit pudgy and inelegant.

Type size tells you what is more and less important (quantitative) on the Voyager map. The title and story text are the largest, grid degrees and ocean names the smallest.

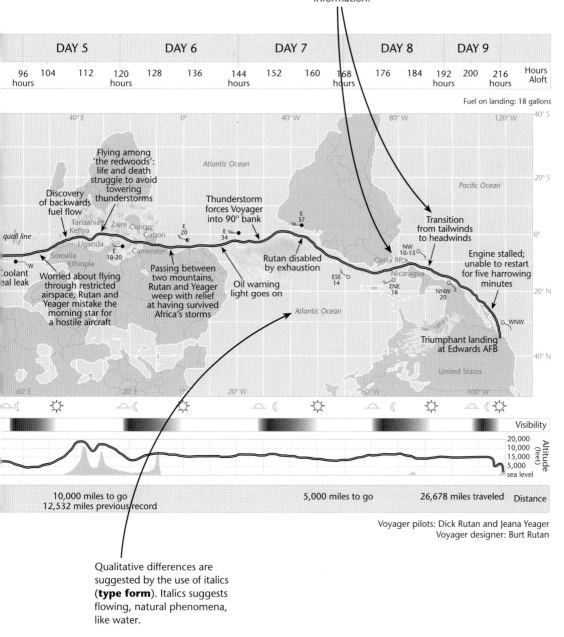

Quantitative differences are suggested by the use of value (**type form**). Gray type is used for less important information, black type for more important information.

DAY 5		DAY 6		DAY 7		DAY 8	DAY 9		

96 hours	104	112	120 hours	128	136	144 hours	152	160	168 hours	176	184	192 hours	200	216 hours	Hours Aloft

Fuel on landing: 18 gallons

40° E   0°   40° W   80° W   120° W   40° S

Atlantic Ocean

Pacific Ocean   20° S

Flying among 'the redwoods': life and death struggle to avoid towering thunderstorms

Discovery of backwards fuel flow

Thunderstorm forces Voyager into 90° bank

Transition from tailwinds to headwinds

E 37

quall line

Tanzania Kenya   Zaire Congo Gabon   E 20   E 34   E

0°

Uganda   Cameroon   E 10-20   NW 10-15   Costa Rica   Engine stalled; unable to restart for five harrowing minutes

Somalia Ethiopia   Rutan disabled by exhaustion   ESE 14   Nicaragua   ENE 18   NNW 20   20° N

W   Coolant eal leak   Worried about flying through restricted airspace, Rutan and Yeager mistake the morning star for a hostile aircraft   Passing between two mountains, Rutan and Yeager weep with relief at having survived Africa's storms   Oil warning light goes on   WNW

Atlantic Ocean   Triumphant landing at Edwards AFB

40° N

United States

60° E   20° E   0°   20° W   60° W   100° W

Visibility

20,000
10,000
15,000
5,000
sea level

Altitude (feet)

10,000 miles to go
12,532 miles previous record

5,000 miles to go   26,678 miles traveled   Distance

Voyager pilots: Dick Rutan and Jeana Yeager
Voyager designer: Burt Rutan

Qualitative differences are suggested by the use of italics (**type form**). Italics suggests flowing, natural phenomena, like water.

Arranging Type on Maps

Effective type placement clarifies the relationship between a label and the symbol (point, line, area) to which it refers. GIS applications may offer automated feature labeling, which normally follows traditional type placement rules. Always evaluate any automated type placement for clarity and legibility.

Labeling Point Data

When labeling point symbols on a map, start at the center of the map and work outward. For each symbol, follow these priorities, where 1 is best, 8 is worst.

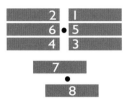

Show characteristics of the labeled location with type placement.

Label ports and harbor towns on the sea
Label inland towns on the land
Label land features on land; water features on water
Label towns on the side of the river on which they are located
Align type to grid (latitude) if grid is included

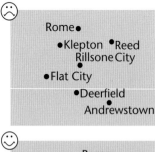

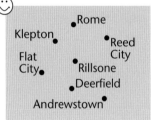

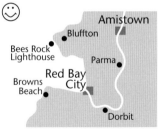

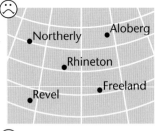

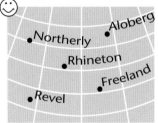

Labeling Line Data

Curve or slant type, following symbol
Keep type above symbol if possible
Keep type as horizontal as possible for ease of
    reading

Never place type upside-down
With vertical type, place the first letter of the
    label at the bottom
Repeat rather than stretch type

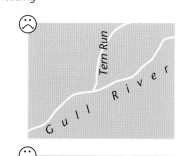

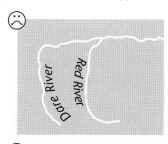

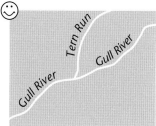

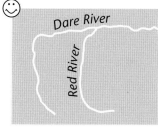

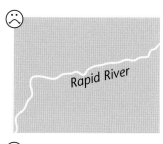

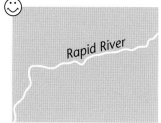

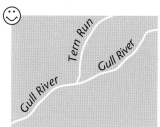

Labeling Area Data

Curve and space type to fit areas. Ensure that the area and the label are clearly associated.

Entire area label should follow a gentle and smooth curve
Keep area labels as horizontal as possible (they are easier to read)
Avoid vertical and upside-down labels (they are harder to read)
Keep labels away from area edges
Avoid hyphenating or breaking up area labels

Distinguish overlapping areas by varying type size, weight, form
Label linear areas like line symbols

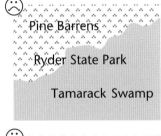

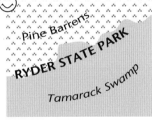

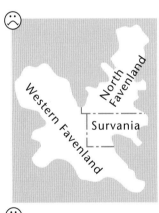

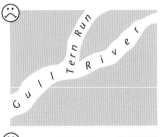

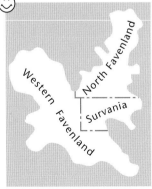

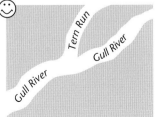

Typographic Minutiae and Maps

Effective type on and around maps requires understanding the basics of typography. Of particular relevance are kerning, letterspacing, line spacing, and alignment. Careful design of type will make your map more functional and beautiful.

Kerning

The combination of Ta and Ve below, when not kerned, look like there is too much space between them. Kerning adjusts the spacing between particular pairs of letters to make them look uniform and less distracting.

Talc Venus
Talc Venus

Kerning is automatically adjusted with most digital type placement algorithms in mapping software, but can also be set by hand.

Evaluate type on or around your map that may need kerning
Kerning is more important for larger type sizes

Letterspacing

Letterspacing, or tracking, changes the spacing between all letters. Normal, positive, and negative letterspacing are shown below.

Talc Venus
Talc Venus
Talc Venus

Avoid negative letterspacing, or take care to avoid hard-to-read, scrunched-up type
Increase letterspacing slightly for a more open, airy feel in a block of text
Increase letterspacing by using upper-case letters to label area features

Line Spacing

Line spacing, or leading, adjusts spacing between lines of text.

This paragraph of text is in Stone Sans Serif 9 point type with 11 point spacing. Normal spacing is set to 120% of the point size of the type.

This paragraph of text is in Stone Sans Serif 9 point type with 9 point spacing. Blocks of text "set solid" look cramped and are harder to read.

This paragraph of text is in Stone Sans

Serif 9 point type with 14 point spacing.

Too much line spacing causes text blocks

to break into what looks like separate lines

of text.

Evaluate type on or around your map that may need line spacing
Maintain a consistent line spacing for similar features labeled on a map
Avoid "set solid" line spacing; instead try a smaller type size
Avoid too much line spacing, as such labels may be misinterpreted as multiple feature labels

Alignment

Avoid left-right justification if it causes distracting spacing problems
Ragged right alignment is the norm, but too much ragged is distracting. Use hyphenation sparingly in text blocks, and avoid it on map labels
Ragged left is more difficult to read in blocks, but use it for map labels referring to symbols to the right of the label

Hawkeye: [looking for maps of the minefield] Why aren't they under "M"?
Radar: Because they're under "B," for "boom."

M.A.S.H. (1973)

"I can't read the names on this [map] because they are all in English."

Christian realized he would have to show his friend how to read a map. "The top is north," he said. "The little circles are town and villages. Blue means rivers and lakes. The thin lines are roads and the thick ones railways."

"There's nothing at all here," said Big Tiger, pointing to one of the many white patches.

"That means it's just desert," Christian explained. "You have to go into the desert to know what it looks like."

Fritz Muhlenweg, *Big Tiger and Christian* (1952)

Rock which does not cover,
Coral reef, detached,
Wreck always partially submerged.

A number of sunken wrecks,
Obstruction of any kind,
Limiting danger line.

Foul ground, discolored water,
Position doubtful,
Existence doubtful.

John Krygier, from map symbol descriptions, *Section O of Chart #1, Nautical Chart Symbols and Abbreviations* (2007)

More...

Most cartography texts devote some space to type on maps. Borden Dent's type chapter in *Cartography: Thematic Map Design* (2008) is probably the best. Typebrewer.org, created by Ben Sheesley and Cindy Brewer, is a great way to visualize and select type for maps.

A very useful book on type is Ellen Lupton, *Thinking with Type: A Critical Guide for Designers, Writers, Editors, and Students* (Princeton Architectural Press, 2004). Why not also peruse Warren Chappell and Robert Bringhurst, *A Short History of the Printed Word* (Hartley and Marks, 2000), and Robert Bringhurst, *Elements of Typographic Style* (Hartley and Marks, 2004); Edward Catich's *The Origin of the Serif* (St. Ambrose University Press, 1991) is a thought-provoking book about letters.

Sources: The type map of Paris is based on one from a chapter Stanley Milgram published in Harold Proshansky, William Ittelson, and Leanne Rivlin, *Environmental Psychology* (Holt, Reinhart, and Winston, 1976). The Waldseemüller map is from Wikipedia (wikipedia.org). The Boylan Heights leaf map is from Denis Wood, *Everything Sings: Maps for a Narrative Atlas* (Siglio Press, 2010). The scribbled-on map of Europe was published in *Life* magazine, December 17, 1914. The Wankers tube map was posted on Geoff Marshall's sillymaps web page; the map was made by an anonymous contributor. The type placement pairs are redrawn from Eduard Imhof, "Positioning Names on Maps," *The American Cartographer,* 2:2, 1975 (pp. 128-144).

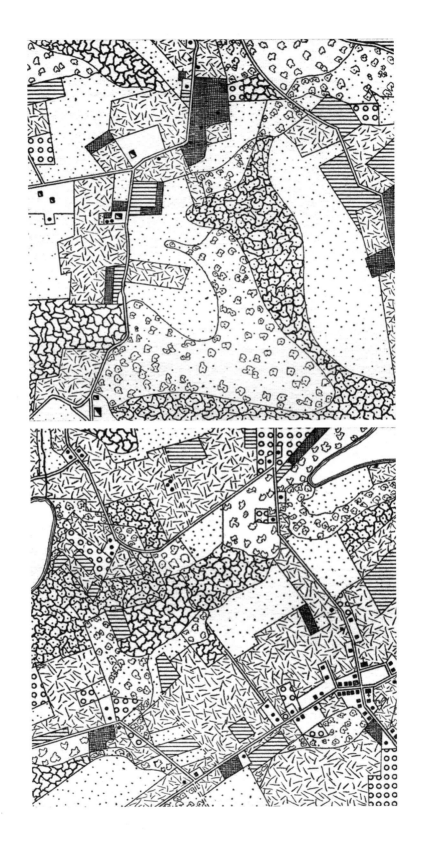

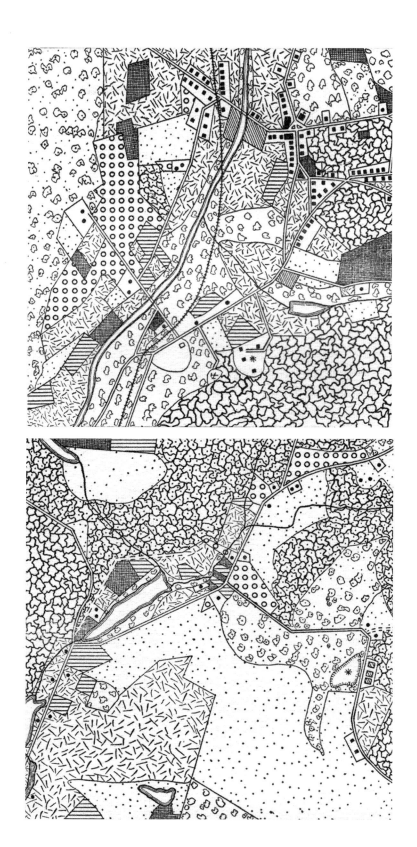

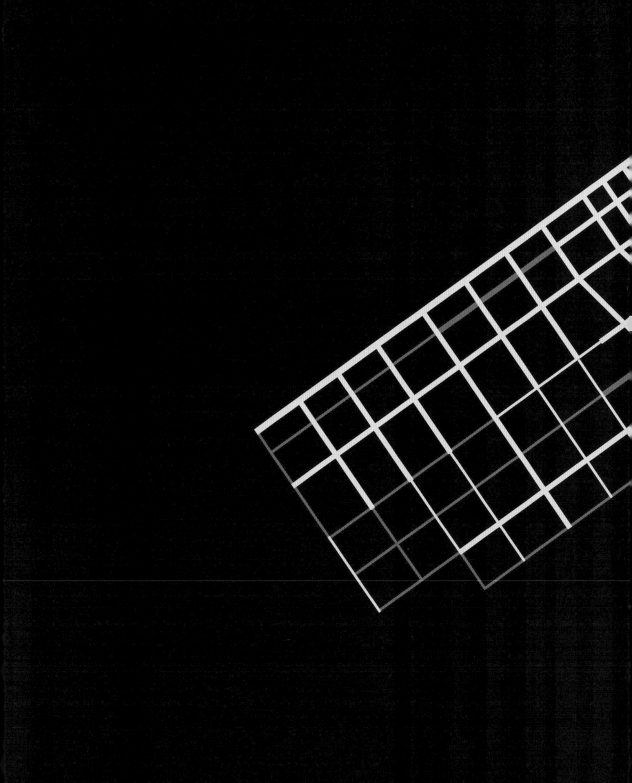

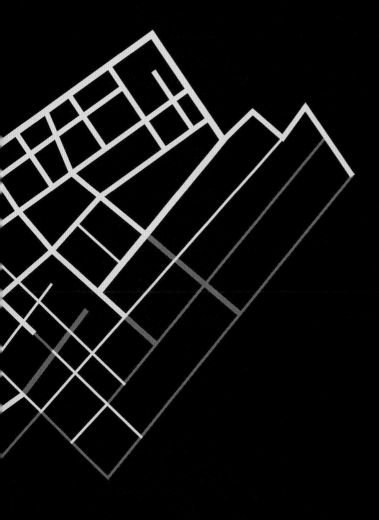

Do you need to move beyond
black and white?

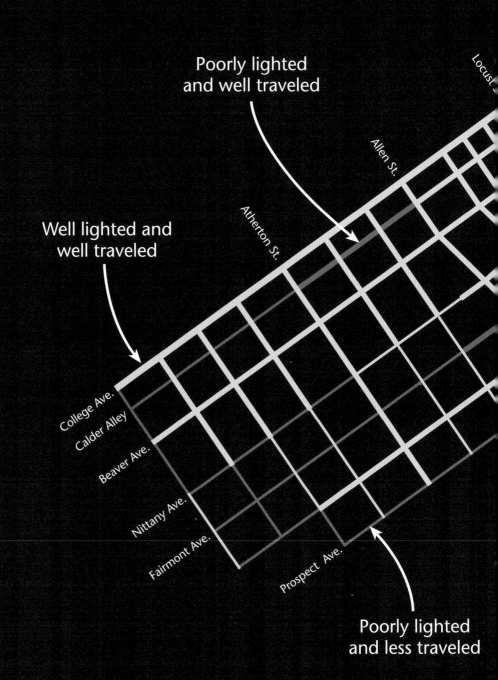

Poorly lighted
and well traveled

Well lighted and
well traveled

Poorly lighted
and less traveled

Locust

Allen St.

Atherton St.

College Ave.

Calder Alley

Beaver Ave.

Nittany Ave.

Fairmont Ave.

Prospect Ave.

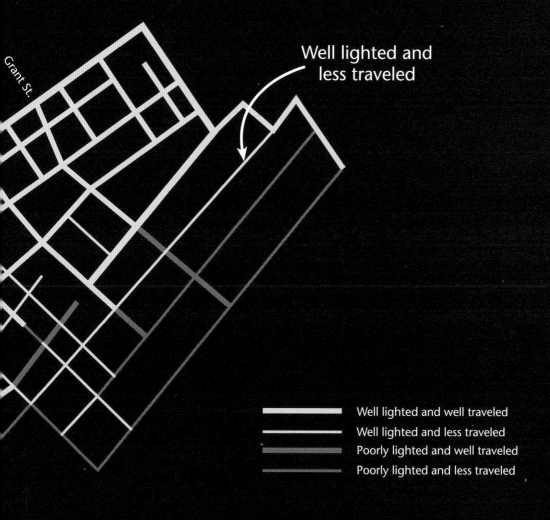

Grant St.

Well lighted and
less traveled

Well lighted and well traveled
Well lighted and less traveled
Poorly lighted and well traveled
Poorly lighted and less traveled

11 Color on Maps

You don't need color to make excellent maps. This "night" map of State College, Pennsylvania, shows how well streets are lighted at night and how many people are around. The map helps people choose a safe route, and the map wouldn't do that any better if color were used.

Thinking about Color on Maps

Color has a huge impact – positive or negative – on the design of your map. When used well, color vastly extends the effectiveness of your map. When used poorly, it easily draws attention away from your data and your goals for the map. Tufte's idea of graphical excellence and the visual variables provide ways to think about appropriate color choices on your map.

Graphical Excellence with Color

"Above all, do no harm" is the adage of Edward Tufte in *Envisioning Information.* Tufte's color Tufteisms, in part drawing from the work of cartographer Eduard Imhof, serve as a guide to excellence of color use on maps.

Graphical excellence is the well-designed presentation of interesting data –
a matter of substance, of statistics, and of design.
Use color with an awareness that adjacent colors perceptually modify each other.
Use strong color for important data in small areas against a muted background.
Use color redundancy to reduce perceptual color shifts and ambiguity.
Use color to distinguish and differentiate features on your map.
Use muted color for less important or background data.
Use color to distinguish order in quantitative data.
Use color to mimic the color of phenomena.
Use muted color over large adjacent areas.
Use color to engage your map's viewers.
Use color palettes found in nature.

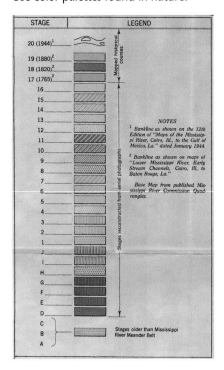

Ancient Courses, Mississippi River Meander Belt, Cape Girardeau, Missouri, to Donaldsonville, Louisiana, Sheet 7

Sheet 7 of the *Ancient Courses of the Mississippi River* map series was published in Harold Fisk's *Geological Investigation of the Alluvial Valley of the Lower Mississippi River* (1944). This spectacular map, expressing engaging data with graphical excellence, reveals changes in the course of the Mississippi River over thousands of years. The map maker differentiates 27 stages of the river. Color (and texture) are used to effectively reveal the tangled knot that is the lower Mississippi. It would be impossible to communicate this complex data without the use of color.

The choice of colors along with the interesting data **engage viewers**, making the subject of fluvial geomorphology seem quite fascinating.

The range of earthy, warm hues used on the map evoke **the phenomena** of ancient river courses.

Color (as well as texture) is chosen to help **distinguish and differentiate** the 27 historical river courses. The challenge is in the sheer number of categories and their complex spatial patterns.

Data of natural phenomena mapped with **natural color palettes** are true to the phenomena, visually engaging, and reveal the complexity of the phenomena.

The **muted** tan background color allows the historical river beds to stand out as the most vital part of this map.

Because the riverbed data are chronological, color value could have been used to **distinguish order.** Instead, the choice was to distinguish qualitative differences, as with the use of color on geologic maps.

Visual Variables and Color

Particular color visual variables suggest particular characteristics of your data. Color hue suggests qualitative differences, color value ordered, quantitative differences. These guidelines apply to point, line, and area map symbols.

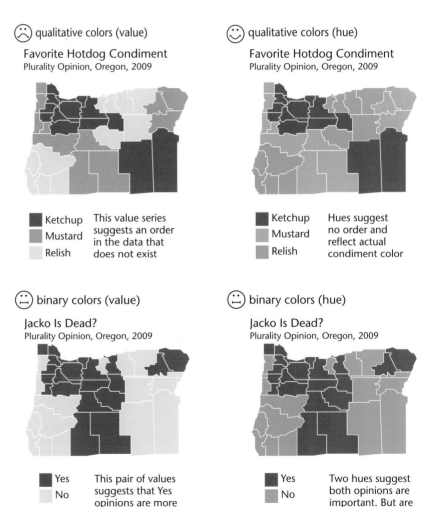

Mapping *Qualitative* Data

qualitative colors (value)

Favorite Hotdog Condiment
Plurality Opinion, Oregon, 2009

Ketchup
Mustard
Relish

This value series suggests an order in the data that does not exist

qualitative colors (hue)

Favorite Hotdog Condiment
Plurality Opinion, Oregon, 2009

Ketchup
Mustard
Relish

Hues suggest no order and reflect actual condiment color

Mapping *Binary* Data

binary colors (value)

Jacko Is Dead?
Plurality Opinion, Oregon, 2009

Yes
No

This pair of values suggests that Yes opinions are more important than No

binary colors (hue)

Jacko Is Dead?
Plurality Opinion, Oregon, 2009

Yes
No

Two hues suggest both opinions are important. But are they?

☹ ordered colors (hue)

Fallen, Can Get Up
per 1000 population, Oregon, 2009

☺ ordered colors (value)

Fallen, Can Get Up
per 1000 population, Oregon, 2009

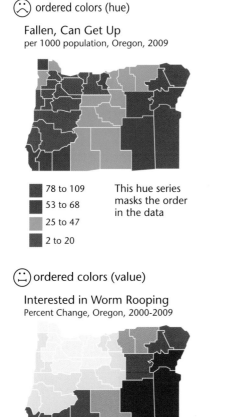

■ 78 to 109	This hue series
■ 53 to 68	masks the order
▨ 25 to 47	in the data
■ 2 to 20	

■ 78 to 109	This value series
■ 53 to 68	reveals the order
▨ 25 to 47	in the data
▢ 2 to 20	

☺ ordered colors (value)

Interested in Worm Rooping
Percent Change, Oregon, 2000-2009

☺ ordered colors (value)

Interested in Worm Rooping
Percent Change, Oregon, 2000-2009

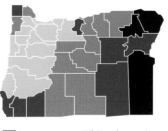

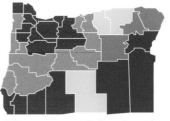

■ 26% to 58%	This value series
■ 1% to 25%	does reveal the
▨ 0 to -99%	ordered data, but...
▢ −100% to − 499%	
▢ −500% to −1000%	

■ 26% to 58%	A diverging value
▨ 1% to 25%	series reveals the
▢ 0 to -99%	diverging data but
▨ −100% to − 499%	may be confusing.
■ −500% to −1000%	

233

Seeing Color on Maps

A diversity of factors, some of which the map maker can control, shape how colors on maps are seen. Attention paid to the light source under which a map is viewed, the surface the map is displayed on, and the basics of human color perception help guide the effective choice of colors when making maps.

Light Source

The colors on a map vary as the light source varies. The same colors will look different when viewing a map

Under daylight, incandescent, or
    fluorescent lighting
As the intensity of the light varies
On a computer monitor, which emits light –
    thus the colors will be brighter and
    more saturated than on paper maps

When selecting colors for a map, consider the conditions under which your map will be viewed.

Low-intensity lighting: use more intense,
    saturated colors
High-intensity lighting: use less intense, less
    saturated colors
Computer monitor: use less intense, less
    saturated colors
Look critically at your map under lighting
    conditions similar to those of your map's
    audience, and adjust the colors to suit

Look at the colors below. Then move to a darker room and look at the colors again. They change. Choose colors for maps that work under appropriate lighting conditions.

Map Surface

The colors on a map vary as the surface the map is displayed on varies. The same colors will look different when viewing a map

On glossy versus matte-surfaced paper
On paper versus projected versus on a computer
    monitor

When selecting colors for a map, consider the effect of the map surface.

Glossy paper will make colors more intense and
    vibrant
Matte paper will make colors less intense and
    dulled
Projectors, depending on the intensity of the
    bulb, may reproduce colors much more or
    less intense than you expect
Computer monitors will make colors intense
    and vibrant, as the color on computer
    monitors is emitted rather than reflected
    (as is the case with paper)
Look critically at your map on the medium the
    map will be presented on, and adjust the
    colors if necessary

Color Dimensions

Our eyes are sensitive to blue, green, and red wavelengths of energy with overlap so we can sense the entire spectrum (red, orange, yellow, green, blue, indigo, violet). One way to think about how people perceive colors is in terms of three dimensions of color perception: hue, value (lightness), and intensity (saturation, chroma).

Hue

Hue is the name for our human experience of particular electromagnetic energy wavelengths. Hues are qualitatively different, thus good for showing qualitative data.

Purple
Yellow
Orange
Blue
Green
Red

Value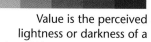

Value is the perceived lightness or darkness of a hue. Values are quantitatively different, thus good for showing quantitative data.

Dark
Darkish
Medium
Lightish
Light

Intensity

Intensity describes the purity of a hue. Intensity is subtle and good for showing binary (yes, no), qualitative, and quantitative data.

Green
Green
Green
Green
Green

Creating Color on Maps

The specification and production of colors are often very different from the way in which we see them. Color specification systems are schemes that organize and help produce different colors. There are many different color specification systems, and map makers will encounter many of them. Three major categories of color specification systems are important: predefined color systems, perceptual color systems, and process color systems.

Predefined Color

Predefined color specification systems are like paint chips from the paint store. Thousands of predefined colors are specified by names or codes. Predefined colors ("spot" colors) are used by some commercial printers and are commonly used when mapping data with set color conventions (such as on geology maps).

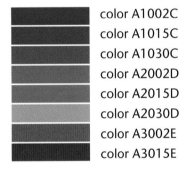

color A1002C
color A1015C
color A1030C
color A2002D
color A2015D
color A2030D
color A3002E
color A3015E

Predefined colors will be converted to another color system when printing on a computer printer or when using a commercial printer who uses process color.

Pantone is a common predefined color system used in mapping.

Perceptual Color

Perceptual color specification systems, such as Munsell, are based on human perceptual abilities. Perceptual tests have produced a set of colors that the average person can differentiate. Thus, no two colors in the Munsell system look exactly alike. The Munsell system consists of a series of color samples, each a single hue with varying value and intensity.

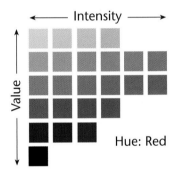

The Munsell system is excellent for selecting appropriate colors for your map, but it will be converted to another system in order to print.

Munsell colors are used as the basis of the ColorBrewer site (colorbrewer.org) created by Cindy Brewer and Mark Harrower. The site converts Munsell colors into other color specification systems so you can easily use the colors in most mapping software.

Process Color: Printing

Process color specification systems use three or four colors to create all other colors. Printed colors typically use the subtractive primaries and rely on reflected light. When you combine cyan (C), magenta (M), and yellow (Y), you produce black – all light is absorbed (subtracted) from your vision. Thus cyan, magenta, and yellow are the subtractive primaries.

Process Color: Monitors

Computer monitors also use three colors to create all other colors. Monitor colors typically create color with the additive primaries and rely on emitted light. Because the light is emitted, the colors are more intense. When you combine red (R), green (G), and blue (B), they add up to pure white. Thus red, green, and blue are the additive primaries.

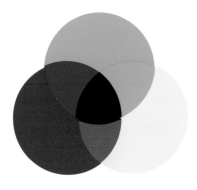

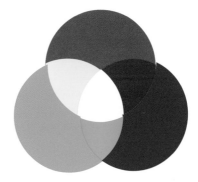

Black is added as a fourth "color" (K, thus CMYK) to avoid the muddy dark brown that is the result of combining cyan, magenta, and yellow.

Subtractive primaries are often used by commercial printers and are common on inkjet computer printers. Different amounts of CMY and K produce thousands of other colors. The CMYK color system should be used for most commercially printed maps.

The hexidecimal color specifications used in HTML (hypertext markup language) are RGB. The first two digits are red, second two digits green, and third two digits blue. 00 is no color and FF is maximum color.

The RGB color system should be used for maps printed with computer printers. RGB will have to be converted into CMYK or predefined color if you plan to print with a commercial printer.

Complicating Color on Maps

The use of color on maps is complex: colors interact with surrounding colors, there are perceptual differences among map viewers, and color has symbolic connotations.

Color Interactions

The appearance of any color on a map depends on surrounding colors. This optical illusion, called simultaneous contrast, makes the left gray dot (below) look slightly darker than the right gray dot (for most people).

Different colors can also look the same, depending on their background. Color subtraction makes the two small squares below look similar.

Yet, they are not.

If the background of a map has varying colors, check that symbols that are supposed to be the same color look the same everywhere on the map.

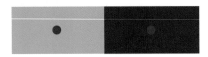

Carefully consider the visual difference between different colors on your map. If you intend for your map to distinguish specific data from other data, use colors that have a high visual difference. Less visual difference is useful if your goal is to suggest less difference between data.

Perceptual Differences

The appearance of color on a map varies, depending on the particular eye-brain system looking at it.

Older map viewers

Benefit from more saturated colors
Have particular difficulty in differentiating shades of blue
Benefit from increasing the type size a bit

Younger map viewers

Like brighter, saturated colors – but not too saturated
Dislike dull, gray, or mixed colors like brown
Perform tasks well with maps that use saturated and unsaturated colors
Understand quantitative, ordered data shown with color value by age 7 or 8

Color-blind viewers typically see red and green as the same. In the U.S., 3% of females and 8% of males are color-blind.

If reds and greens show important differences on your map, a significant number of viewers will not see these differences
Consider using reds and blues or greens and blues instead
Check internet resources for selecting color-blind safe colors

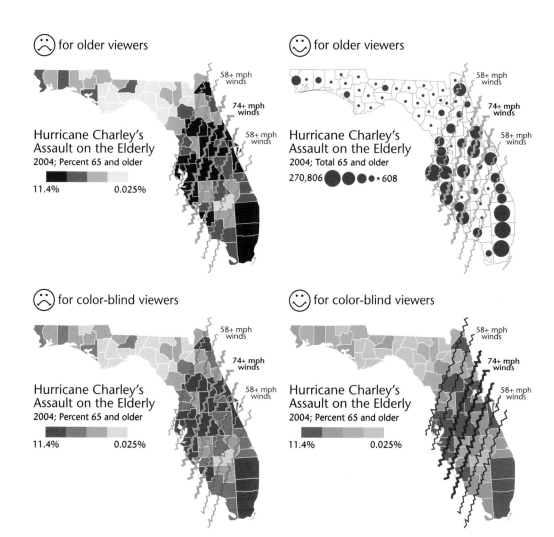

for older viewers

Hurricane Charley's
Assault on the Elderly
2004; Percent 65 and older

11.4%   0.025%

58+ mph
winds

74+ mph
winds

58+ mph
winds

for older viewers

Hurricane Charley's
Assault on the Elderly
2004; Total 65 and older

270,806 ●●●●• 608

58+ mph
winds

74+ mph
winds

58+ mph
winds

for color-blind viewers

Hurricane Charley's
Assault on the Elderly
2004; Percent 65 and older

11.4%   0.025%

58+ mph
winds

74+ mph
winds

58+ mph
winds

for color-blind viewers

Hurricane Charley's
Assault on the Elderly
2004; Percent 65 and older

11.4%   0.025%

58+ mph
winds

74+ mph
winds

58+ mph
winds

Symbolic Color Connotations

Color has symbolic connotations. Such connotations subtly shape viewer reactions and should be guided by your goals for your map. Generic Western cultural color connotations include:

Blue: water, cool, positive numbers, serenity, purity, depth

Green: vegetation, lowlands, forests, youth, spring, nature, peace

Red: warm, important, negative numbers, action, anger, danger, power, warning

Yellow/tan: dry, lack of vegetation, intermediate elevation, heat

Orange: harvest, fall, abundance, fire, attention, action, warning

Brown: landforms (mountains, hills), contours, earthy, dirty, warm

Purple: dignity, royalty, sorrow, despair, richness, elegant

White: purity, clean, faith, illness, life, clarity, absence, light

Black: mystery, strength, heaviness, death, nighttime, presence

Gray: quiet, reserved, sophisticated, controlled, light, bland, dull

Cultural Color Connotations

The symbolic connotations of different colors varies from culture to culture, further complicating the use of color on maps. Check for cultural color connotations if you are mapping for a global audience.

Blue: safe cross-cultural color, because it is the color of the sky, which stands over all peoples

Green: fertility and paganism in Europe, sacred for Muslims, mourning and unhappiness in Asia

Red: Bolsheviks, communists, and other politically left organizations, purity in India

Yellow/tan: peaceful resistance movement associated with Carazon Aquino in Philippines

Orange: pro-Western activists in Ukraine, Protestants in Ireland, sacred Hindu color

Brown: mourning in India, Nazis in West, ceremonial for Australian Aboriginals

Purple: death and crucifixion in Europe, mysticism, prostitution in the Middle East

White: unhappiness in India, mourning in China, royalists and traditionalists in Western

Black: fascists, anarchists, and other extremists in Western world, death, mourning in West

Gray: corporate culture in the West (also blue), dead and dull in Feng Shui

 color conventions

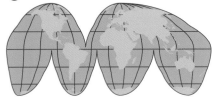

color conventions

Emotion, Experience, and Color on Maps

Color used as mere decoration on maps is an agnostic sin. With a critical eye cast upon the conventions of traditional maps, Margaret Pearce and Michael Hermann designed and produced a narrative map of the travels and experiences of Samuel de Champlain in Canada. Their goal, to design a map expressing the emotions, voices, and multiple experiences of Champlain, his men, and indigenous peoples: interesting data and complex ideas presented with clarity and intelligence. Graphical excellence with color.

Color and Experience

Different voices and experiences lead to different maps of the same place, issue, or phenomenon, or you can embed them in one map. Pearce and Hermann use color hue (right) to map the multiple voices and experiences in the Champlain narrative. Champlain in blue, indigenous people in green, and the voice of the map maker, from a future time and place, in gray.

Color and Emotion

Color is emotive: angry red, calm green, depressed gray, happy yellow. Pearce and Hermann use color hue to express shifting emotions from panel to panel on the Champlain map. Below, Champlain learns of an assassination plot against him, and the colors differentiate the different voices and shifting emotions of Champlain and the assassins.

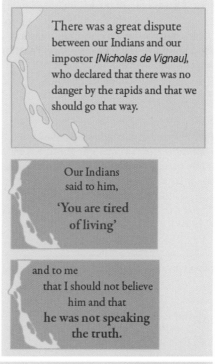

There was a great dispute between our Indians and our impostor *[Nicholas de Vignau]*, who declared that there was no danger by the rapids and that we should go that way.

Our Indians said to him, 'You are tired of living'

and to me that I should not believe him and that he was not speaking the truth.

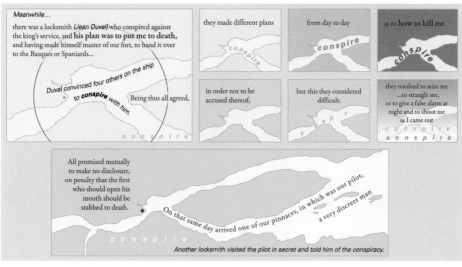

Meanwhile...
there was a locksmith *[Jean Duval]* who conspired against the king's service, and his plan was to put me to death, and having made himself master of our fort, to hand it over to the Basqués or Spaniards...

Duval convinced four others on the ship to *conspire* with him. Being thus all agreed,

they made different plans

from day to day

as to how to kill me

conspire

in order not to be accused thereof,

but this they considered difficult.

they resolved to seize me ...to strangle me, or to give a false alarm at night and to shoot me as I came out

All promised mutually to make no disclosure, on penalty that the first who should open his mouth should be stabbed to death.

On that same day arrived one of our pinnaces, in which was our pilot, a very discreet man

Another locksmith visited the pilot in secret and told him of the conspiracy.

If we don't want to see the map of Central America covered in a sea of red, eventually lapping at our own borders, we must act now.

Ronald Reagan (1986)

Fozzie Bear: Kermit, where are we?
Kermit the Frog: [looking at a map] Well, let's see. We're just traveling down this little black line here, and uh, just crossed that little red line over here.
Fozzie Bear: [takes his eyes off the road to focus on the map] Look, why don't we just take that little blue line, huh?
Kermit the Frog: We can't take that. That's a river.
Fozzie Bear: Oh. I knew that.
Kermit the Frog: Yeah, sure.

The Muppet Movie (1976)

I do not advance that the face of our country would change if the maps which Philadelphia sends forth all over the Union were more decently colored, but certainly it would indicate that the Graces were more frequently at home on the banks of our lovely rivers, if the engravers were able to sell their maps less boisterously painted and not as they are now, each county of each state in flaming red, bright yellow, or a flagrant orange dye, arrayed like the cover produced by the united efforts of a quilting match. When I once complained of this barbarous offensive coloring of maps, the geographer assured me that he would not sell them unless bedaubed in this way; "for," said he, "the greatest number of the large maps are not sold for any purpose of utility, but to ornament the walls of barrooms. My agents write continually to me to color high." This reason was given me by one of the first geographers of the United States, who has himself a perfectly correct idea of the tasteful coloring of maps.

Francis Lieber, "On Hipponomastics: A Letter to Pierce M. Butler," *Southern Literary Messenger,* 3:5 (1837).
   Thanks to Penny Richards for this quote.

More...

Cindy Brewer's research on color for maps has been integrated into the very useful colorbrewer.org website. It is a great way to select effective color for maps.

A great article on natural color maps is Tom Patterson and Nathaniel Vaughn Kelso's "Hal Shelton Revisited: Designing and Producing Natural-Color Maps with Satellite Land Cover Data," available with a bunch of other cool stuff at the shadedrelief.com website.

Color Oracle is a very useful free software application that simulates three types of color blindness on your computer screen (colororacle.cartography.ch).

Edward Tufte engages color in all of his books, including a whole chapter on "Color and Information" in *Envisioning Information* (Graphics Press, 1990).

For some solid background on the history and theory of color: John Gage, *Color and Culture: Practice and Meaning from Antiquity to Abstraction* (University of California Press, 1999) and *Color and Meaning: Art, Science, and Symbolism* (University of California Press, 2000); and Charles A. Riley II, *Color Codes: Modern Theories of Color in Philosophy, Painting and Architecture, Literature, Music, and Psychology* (University Press of New England, 1995).

Sources: The State College night map was redrawn from the original created in the Deasy GeoGraphics Lab (now the Gould Center) at Penn State. The *Ancient Courses, Mississippi River Meander Belt* maps are available in digital form from the *Lower and Middle Mississippi Valley Engineering Geology Mapping Program* (lmvmapping.erdc.usace.army.mil). The idea for the hurricane maps came from a map published in the *News & Observer* (Raleigh-Durham, NC) on August 19, 2004. Excerpts of Pearce and Hermann's map "They Would Not Take Me There: People, Places and Stories from Champlain's Travels in Canada, 1603-1616" are used by permission. See Margaret Pearce and Michael Hermann, "Mapping Champlain's Travels: Restorative Techniques for Historical Cartography." 2010. *Cartographica* 45:1, pp. 32-46. The map is available from the Canadian-American Center at the University of Maine (www.umaine.edu/canam).

Maps between Each Chapter

A series of mysterious and enchanting maps is placed between each chapter. The maps were modified to remove distracting details; please don't use them as real data.

"Unincorporated Hamlets in Wisconsin." Glenn Trewartha. 1943. "The Unincorporated Hamlet." *Annals of the Association of American Geographers* 33:1, pp. 32-81. (between introduction and chapter 1)

"Map of part of Naranjo." Sylvanus G. Morley. 1909. "The Inscriptions of Naranjo, Northern Guatemala." *American Anthropologist, New Series,* 11:4, pp. 543-562. "The Yorkshire, Derbyshire, and Nottinghamshire Coal-Field." Rodwell Jones. 1921. "Commodity Maps." *Economica,* 3, pp. 246-258. (between chapters 1 and 2)

"Urban Heat Islands." Warner Terjung and Stella Louie. 1973. "Solar Radiation and Urban Heat Islands." *Annals of the Association of American Geographers* 63:2, pp. 181-207. (between chapters 2 and 3)

"Passenger and Freight Railroad Networks,1946." Charles Hitchcock. 1946. "Westchester-Fairfield: Proposed Site for the Permanent Seat of the United Nations." *Geographical Review* 36:3, pp. 351-397. "The Distribution of Population in Romania (1930)." J.M. Houston. 1953. *A Social Geography of Europe.* London: Duckworth. (between chapters 3 and 4)

"Meanders in Anatolian Rivers." Richard Russell. 1954. "Alluvial Morphology of Anatolian Rivers." *Annals of the Association of American Geographers* 44:4, pp. 363-391. (between chapters 4 and 5)

"Percentage of the Sections Entered under the Timber Culture Act" and "Percentage of Public Domain Entered before 1872" (Nebraska, Kansas, Missouri, Illinois, and Minnesota). C. Barron McIntosh. 1975. "Use and Abuse of the Timber Culture Act." *Annals of the Association of American Geographers* 65:3, pp. 347-362. "Movement of Summer Vacationists within Luce County Michigan, 1929." George Deasy. 1944. "The Tourist Industry in a 'North Woods' County." *Economic Geography* 25:4, pp. 240-259. (between chapters 5 and 6)

"Plan of Typical House - Nangodi, Ghana." John Hunter. 1967. "The Social Roots of Dispersed Settlement in Northern Ghana." *Annals of the Association of American Geographers* 57:2, pp. 338-349. "Generalized Winter Storm Patterns" and "Meandering River." Robert Ward. 1914. "The Weather Element in American Climates." *Annals of the Association of American Geographers* 4, pp. 3-54. (between chapters 6 and 7)

"Cars of Malting Barley By County of Origin, 1939-40." John C. Weaver. 1944. "United States Malting Barley Production." *Annals of the Association of American Geographers* 34:2. pp. 97-131. "Rockford, Illinois." John Alexander. 1952. "Rockford, Illinois: A Medium-Sized Manufacturing City." *Annals of the Association of American Geographers* 42:1, pp. 1-23. (between chapters 7 and 8)

"Results of Traffic Census, 1913, for South-western London." Aston Webb. 1918. "The London Society's Map, with Its Proposals for the Improvement of London." *The Geographical Journal,* 51:5, pp. 273-287. "Tortugas." Alexander Agassiz. 1885. "Explorations of the Surface Fauna of the Gulf Stream, under the Auspices of the United States Coast Survey: The Tortugas and Florida Reefs." *Memoirs of the American Academy of Arts and Sciences, New Series* 11:2, No. I, pp. 107-133. (between chapters 8 and 9)

"The Location of Agricultural Production after Von Thunen." Andreas Grotewold. 1959. "Von Thunen in Retrospect." *Economic Geography* 35:4, pp. 346-355. "Losses: The Great Chicago Fire." *The American Architect and Building News,* March 11, 1905. (between chapters 9 and 10)

"Chorography of Southern New England – Selected Areas." Preston James. 1929. "The Blackstone Valley." *Annals of the Association of American Geographers* 19:2, pp. 67-92. (between chapters 10 and 11)

"Computerkarte Regionaldatei." Ulrich Stoye. 1975. "Herstellung Mehrfarbiger Themakarten auf der Grundlage von Schnelldrukerkarten." *International Yearbook of Cartography* 1975, pp. 158-164. (after chapter 11).

A Note to the Users of *Making Maps*

Making Maps is not like other map and cartography texts. It's concise, graphic (as befits the subject), focused on the map-making process, includes specific guidelines to help you design better maps, asks you to think, and shows maps that matter in the real world, engaged with conflict, human curiosity, politics, discovery, and controversy. We are aware of the ever-increasing price of textbooks. As such, we have limited the use of color in the book to keep down its cost. Many map design concepts can be imparted without the use of color; where color is necessary, it is used.

Making Maps was designed for a smart, general audience who want to understand and engage in map making. As such, we have provided substantive examples, left out superfluous jargon, and included guidelines that work regardless of the map-making tools you are using. Resources included at the end of each chapter and the book's blog (makingmaps.net) will lead you to the wonderful abundance of additional information about maps and map making available in books and on the internet.

Making Maps was also designed for use in courses on mapping, cartography, and GIS at the introductory or advanced level. It can serve as the sole text in a course or supplement another text. Given the diverse approaches to maps and mapping in courses in and outside of geography programs (where courses on maps are usually taught), we attempt to provide key concepts and good examples relevant to just about any way one might choose to teach about maps. Course instructors will undoubtedly expand upon the content covered in *Making Maps* in lecture and/or laboratory sessions (a good reason for students to show up for class) and fit the text into their vision of maps and map making.

This book, like any book, reflects the personality, quirks, and intellectual interests of its authors. For better or worse, we just didn't think the world needed another boring text on something as interesting as maps.

Materials Reproduced in *Making Maps*

The authors have created almost all of the illustrations in this book. Reproduced illustrations are indicated in the sources section at the end of each chapter. Every attempt has been made to secure the copyright to material reproduced in this book.

Acknowledgments

Denis Wood: I need once again to thank Christine Baukus and Irv Coats for their continuing support and John Krygier for inviting me to join him on what has turned out to be a really cool trip. Chandler was a lot of fun to work with too!

John Krygier: Feedback from readers of the first edition of *Making Maps* has made the second edition much better. That and Denis yelling and shaking his fist at half the stuff in the first edition of the book. Thanks also to Margaret Pearce and Jeremy Crampton for their great ideas and feedback. A big thanks to my family: Patti, John Riley, and Annabelle, and my parents, Jim and Charlotte, who always think what I do is cool. Thanks to Ohio Wesleyan University for help defraying costs associated with the book.

Check out the book's blog for more stuff related to this book: makingmaps.net

Book design and production by John Krygier.

Illustrations by Chandler Wood.

About the Authors

John Krygier teaches in the Department of Geology and Geography at Ohio Wesleyan University, with teaching and research specializations in cartography, geographic information systems (GIS), as well as environmental and human geography. He has made scads of maps and published on map design, educational technology, cultural geography, multimedia in cartography, planning, the history of cartography, and participatory geographic information systems. He has a master's degree from the University of Wisconsin, where he worked with David Woodward and a PhD from The Pennsylvania State University, where he worked with Alan MacEachren. See krygier.owu.edu for more information.

Denis Wood holds a PhD in geography from Clark University, where he studied map making under George McCleary. He curated the award-winning *Power of Maps* exhibition for the Smithsonian and writes widely about maps. Recent books include *The Natures of Maps* (University of Chicago Press, 2008), *Rethinking the Power of Maps* (Guilford Press, 2010), and *Everything Sings: Maps for a Narrative Atlas* (Siglio, 2011). A former professor of design at North Carolina State University, Wood is currently an independent scholar living in Raleigh, North Carolina. See deniswood.net for more information.

Chandler Wood is an illustrator in Los Angeles, California. Among other things, he illustrates Another LA Story for the *LA Weekly,* and regularly exhibits his work around Los Angeles.

Index